BESTSELLING BOOK SERIES

Lean Six Sigma For Dummies®

The Key Principles of Lean Six Sigma

In a nutshell, here are the key principles of Lean Six Sigma to bear in mind:

- Focus on the customer
- Identify and understand how the work gets done (the *value stream*, explained in Chapter 5)
- Manage, improve, and smooth the process flow
- Remove Non-Value-Added steps and waste (head to Chapter 9 for more on this)
- Manage by fact and reduce variation
- Involve and equip the people in the process (Chapter 12 explores the people issues)
- Undertake improvement activity in a systematic way

Understanding Value

In the customers' eyes, *value* is what they're willing to pay for:

The right products and services

At the right time

At the right price

At the right quality

For a step to be value-added, it must meet the following three criteria:

1. The customer has to care about the step
2. The step must either physically change the product or service in some way, or be an essential prerequisite for another step
3. The step must be actioned 'right first time'

Try to remove those steps that don't meet these criteria, but recognise that you may want to retain some non-value-added steps, perhaps for regulatory or financial reasons, for example.

You need to identify and understand the value stream and eliminate waste and non-value-added steps. As little as 10 to 15 per cent of process steps add value, often representing only 1 per cent of the total process time.

Chapter 9 is all about identifying value-added steps and eliminating waste.

Lean Six Sigma For Dummies®

DMAIC at a Glance

To undertake improvement activity in a systematic way, you need DMAIC:

- Define: Projects start with a problem that needs solving. Make sure everyone involved knows their role, why you're doing the project, and what you're trying to achieve with the project.
- Measure: The work you've done in the Define stage is based on what you think the problem is. During the Measure stage you need to clarify things by seeing how the work gets done and how well.
- Analyse: Now you know what's happening, it's time to find out why, but don't jump to conclusions. Manage by fact to check out the possible causes and get to the root cause.
- Improve: Okay; you know about the process and the problem, and the Improve stage is where you need to find a way to address the root cause, so come up with some ideas, select the best one, and test it out.
- Control: You need to ensure you achieve and hold the gain you're looking for. Putting a control plan in place is vital to ensure that the process is carried out consistently.

Turn to Chapter 2 for lots more information about DMAIC.

Tackling Waste

You can improve process flow in a number of ways, including by reducing waste. The seven categories of waste are sometimes identified by the acronym Tim Wood:

- Transportation: Moving materials and output unnecessarily.
- Inventory: Overproduction resulting in too much stock.
- Motion: Inappropriate siting of teams or equipment.
- Waiting: Equipment failure, for example, which causes delays.
- Over-Processing: Performing unnecessary processing steps.
- Over-Production: Producing more stock or producing it earlier than needed.
- Defects: Dealing with rework.

Find out about reducing waste in Chapter 9.

Lean Six Sigma FOR DUMMIES®

by John Morgan and Martin Brenig-Jones

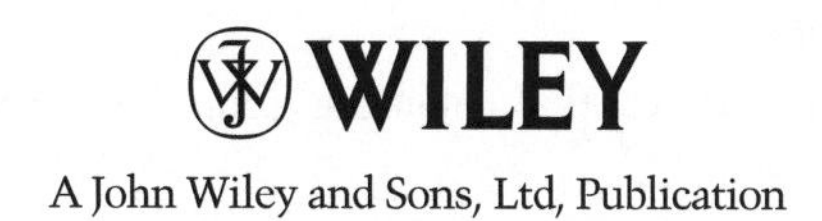

A John Wiley and Sons, Ltd, Publication

Lean Six Sigma For Dummies®

Published by
John Wiley & Sons, Ltd
The Atrium
Southern Gate
Chichester
West Sussex
PO19 8SQ
England

E-mail (for orders and customer service enquires): cs-books@wiley.co.uk

Visit our Home Page on www.wiley.com

Published by John Wiley & Sons, Ltd, Chichester, West Sussex

For general information on our other products and services, please contact our Customer Care Department within the U.S. at 800-762-2974, outside the U.S. at 317-572-3993, or fax 317-572-4002.

For technical support, please visit www.wiley.com/techsupport.

Wiley also publishes its books in a variety of electronic formats. Some content that appears in print may not be available in electronic books.

British Library Cataloguing in Publication Data: A catalogue record for this book is available from the British Library

ISBN: 978-0-470-75626-3

Printed and bound in Great Britain by TJ International Ltd

10 9 8 7 6 5 4 3 2 1

About the Authors

John Morgan is the author of several books, including *The Lean Six Sigma Improvement Journey*, and is co-author of *SPC in the Office*. His experience has led to him being interviewed on BBC Radio 4 about the potential of Lean and Six Sigma in the UK, especially in the public sector and National Health Service.

John has been a Director of the Lean Six Sigma specialists Catalyst Consulting for over 10 years, and much of their highly acclaimed material has been created by him, including tailored work for companies such as General Electric, Saint-Gobain Glass, and British Telecom. John's primary responsibilities are in the areas of product design and development. In addition to his role with Catalyst, John also jointly heads the Lean Six Sigma Academy on behalf of the British Quality Foundation.

A Chartered Insurer and Fellow of the Chartered Institute of Insurance, John's early career background was in Aviation Insurance and Reinsurance. He first started to apply Lean Six Sigma techniques in his role of Customer Services Director for an American insurance company before joining Catalyst.

Martin Brenig-Jones is a Director at Catalyst and has trained and coached over 1,000 people in Lean Six Sigma and business improvement techniques working mainly across Europe and occasionally in the US and Asia. He has worked with organisations in a diverse range of sectors including IT, transport, computer manufacturing, local government, police, health, aerospace, rail, telecoms, and financial services.

Prior to joining Catalyst in 2000, Martin was Head of Quality at BT with responsibility for Quality and Business Excellence across the group. Martin is a member of LRQA's General Technical Committee.

Martin studied Electronic Engineering at the University of Liverpool and has a Postgraduate Diploma in Management. He is a member of the Institute of Engineering and Technology, and in his earlier career he worked in telecommunications, software, and systems development. Martin also gets involved in local education as a school governor which keeps his feet firmly on the ground.

Authors' Acknowledgements

Writing a book tends to take a fair bit of concentration, time, and effort, so there's a general thank you to my wife Margaret, family, and friends for putting up with that.

In particular though, I'd like to thank my long-time friend and colleague Jo Ballard for her help with typing, support, and input to the book, both in keeping me on track, and in helping me get the grammar, or the scanning of the words in a sentence, right.

The team at Wiley have provided great support, too, especially Rachael Chilvers, Nicole Hermitage, and Colette Holden. Finally, everyone at Catalyst has helped in some way, too, especially through their thoughts and ideas that in part have helped shape my thinking over the years.

JM

I'd particularly like to thank my wife, Di, who is the best wife, friend, mother, and teacher in the world. She lives with someone who spends far too much time doing this work stuff and who ought to keep his office at home a lot tidier by actually practicing 5S. I'd also like to thank my four children, Jo, Laurence, Alex, and Oliver for being so patient with the bod who is always seeing 'process improvement opportunities' – particularly annoying when on holiday. Finally I must thank everyone at Catalyst and my clients who have given me the experience that I have tried to distil into this book.

MB-J

Publisher's Acknowledgements

We're proud of this book; please send us your comments through our Dummies online registration form located at www.dummies.com/register/.

Some of the people who helped bring this book to market include the following:

Acquisitions, Editorial, and Media Development

Project Editor: Rachael Chilvers

Content Editor: Jo Theedom

Development Editor: Colette Holden

Copy Editor: Kate O'Leary

Proofreader: Martin Key

Commissioning Editor: Nicole Hermitage

Executive Project Editor: Daniel Mersey

Cover Photos: Tetra Images/Alamy

Cartoons: Ed McLachlan

Composition Services

Project Coordinator: Lynsey Stanford

Layout and Graphics: Nikki Gately, Christin Swinford, Ronald Terry

Indexer: Claudia Bourbeau

Contents at a Glance

Table of Contents

Introduction

Lean Six Sigma provides a rigorous and structured approach to help manage and improve performance. It helps you use the right tools, in the right place, and in the right way, not just in improvement, but also in your day-to-day management of activities. Lean Six Sigma really is about getting key principles and concepts into the DNA and lifeblood of your organisation so that it becomes a natural part of how you do things.

This book seeks to help managers and team leaders better understand their role and improve organisational efficiency and effectiveness.

If you want to change outcomes, you need to realise that outcomes are the result of systems. Not the computer systems, but the way people work together and interact. And these systems are the product of how people think and behave. So, if you want to change outcomes, you have to change your systems, and to do that, you have to change your thinking. Albert Einstein summed up the need for different thinking very well:

> *The significant problems we face cannot be solved by the same level of thinking which caused them.*

Lean Six Sigma thinking is *not* about asset stripping and 'making do'. Instead, this approach focuses on doing the right things right, so that you really do add value for the customer and make your organisation effective and efficient.

DMAIC (Define, Measure, Analyse, Improve, and Control) is the Lean Six Sigma method for improving existing processes, and it provides an ideal way to help you in your quest for continuous improvement.

About This Book

This book makes Lean Six Sigma easy to understand and apply. We wrote it because we feel that Lean Six Sigma can help organisations of all shapes and sizes, both private and public, improve their performance in meeting their customers' requirements.

In particular, we wanted to draw out the role of the manager and provide a collection of concepts, tools, and techniques to help him or her carry out the job more effectively. We also wanted to demonstrate the genuine synergy achieved through the combination of Lean and Six Sigma. For some reason unknown to the authors, a few people feel they can use only Lean or Six Sigma, but not both. How wrong they are!

In this book you can discover how to create genuine synergy by applying the principles of Lean and Six Sigma together in your day-to-day operations and activities.

Conventions Used in This Book

Lean Six Sigma uses a whole range of acronyms, but when they first appear we describe them in full and then use them in their abbreviated form.

We use some statistical concepts and language, but minimise these in order to demonstrate the range of straightforward tools and techniques that can be applied in everyday activity, as well as in improvement projects.

If you'd like more detail and information about some of the statistical aspects, check out *Statistics For Dummies* by Deborah Rumsey, *Six Sigma For Dummies* by Craig Gygi, Neil DeCarlo, and Bruce Williams, and *Lean For Dummies* by Natalie J. Sayer and Bruce Williams, all published by Wiley.

Web addresses appear in `monofont` and new terms in *italics*.

Foolish Assumptions

In Lean Six Sigma, avoiding the tendency for people, and managers in particular, to jump to conclusions and make assumptions about things is crucial. Lean Six Sigma really is about managing by fact. Despite that, we've made some assumptions about why you may have bought this book:

- You're contemplating applying Lean Six Sigma in your business or organisation, and you need to understand what you're getting yourself into.
- Your business is implementing Lean Six Sigma and you need to get up to speed. Perhaps you've been lined up to participate in the programme in some way.
- Your business has already implemented either Lean or Six Sigma and you're intrigued by what you might be missing.

- You're considering a career or job change and feel that your CV will look much better if you can somehow incorporate Lean or Six Sigma into it.
- You're a student in business, operations, or industrial engineering, for example, and you realise that Lean Six Sigma could help shape your future.

We also assume that you realise that Lean Six Sigma demands a rigorous and structured approach to understanding how your work gets done and how well it gets done, and how to go about the improvement of your processes.

How This Book Is Organised

We break this book into five separate parts. Each is written as a stand-alone section, enabling you to move about the book and delve into a given topic without necessarily having to read the preceding material first.

Naturally, with a topic such as Lean Six Sigma, a lot of interrelationship exists between the chapters and where this occurs we provide cross-references so you can tie everything together.

Part I: Lean Six Sigma Basics

In this part we get back to basics, providing an overview of what Lean thinking and Six Sigma really mean, as well as some observations about what they don't mean!

We highlight the synergy created by merging the two disciplines into Lean Six Sigma and provide an overview of the key principles underpinning the approach.

This part explains just what a 'sigma' is and introduces the commonly used process improvement method known as DMAIC (Define, Measure, Analyse, Improve, and Control).

Part II: Working with Lean Six Sigma

Throughout the book, we encourage you to keep asking yourself how and why things are done. What's the purpose of your products and services and the processes that support them?

Ideally, things are done in order to meet the requirements of your customers, but you need to know who they are, or who they might be.

This part focuses on identifying your various, and often quite different, customers, seeing how you can determine their requirements, and showing how to use this information to form the basis of the measurement set for your processes. In doing so, you need to take a brief look at some process basics, too. By drawing a process map you can see what the process really looks like, and understand who does what, when, where, and why.

In essence, you're developing a picture of your customers and the processes that seek to meet their requirements.

Part III: Assessing Performance

In this part, we look to see how well the work gets done. Are you meeting your customers' requirements in the most effective and efficient way?

Managing by fact is a key principle in Lean Six Sigma, so having good data is vital. Data collection is a process in itself, and we present a five-step approach to ensuring you have an appropriate plan in place.

When you have your data, you need to decide how best to present and interpret it. We cover the importance of control charts to help you identify process variation so that you know when to take action and when not to.

We also look at developing an appropriately balanced set of measures that help you understand what influences and affects your results.

Part IV: Improving the Processes

A variety of tools and techniques come together to help you reduce waste and the time it takes to do things. We provide a common definition for what people mean by 'value-added' and 'non-value-added', and look at the importance of identifying and tackling bottlenecks in your processes.

In essence, we look at how to improve the process flow so that things are actioned in less time and with less effort. In doing so, we cover a number of concepts, including 'pull not push', the power of prevention, and the importance of people issues in ensuring successful change.

Part V: The Part of Tens

In these short chapters we provide information on some of the excellent Lean Six Sigma resources available. We also provide a collection of lists that include best practices, common mistakes and where to go for help.

Icons Used In This Book

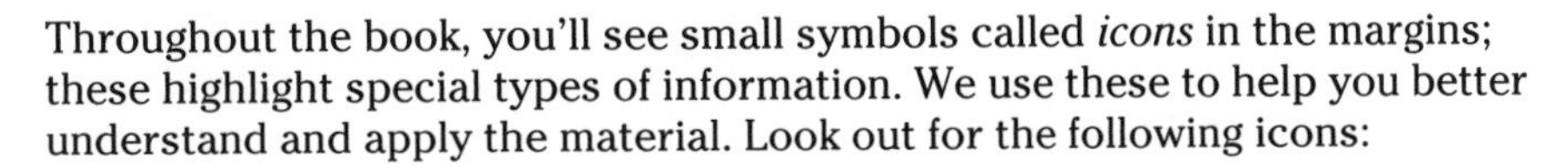

Throughout the book, you'll see small symbols called *icons* in the margins; these highlight special types of information. We use these to help you better understand and apply the material. Look out for the following icons:

The key highlights an essential component of Lean Six Sigma.

Bear these important points in mind as you get to grips with Lean Six Sigma.

Keep your eyes on the target to find tips and tricks we share to help you make the most of Lean Six Sigma.

Throughout this book we share true stories of how different companies have implemented Lean Six Sigma to improve their processes. We also share true stories of when things go wrong so you learn from others' mistakes.

This icon highlights potential pitfalls to avoid.

Where to go From Here

In theory, when you read you begin with ABC, and when you sing you begin with doh-ray-me (apologies to Julie Andrews). But with a *For Dummies* book you can begin where you like. Each part and, indeed, each chapter, is self-contained, which means you can start with whichever parts or chapters interest you the most.

That said, if you're new to the topic starting at the beginning makes sense. Either way, lots of cross-referencing throughout the book helps you see how things fit together and to put them in the right context.

Part I
Lean Six Sigma Basics

'Now we'll see who comes out on top — the traditional way or the new way.'

In this part . . .

Here you meet the basics, as we provide an overview of what Lean Thinking and Six Sigma mean, as well as some observations about what they don't mean!

This part highlights the synergy created by merging the two disciplines into Lean Six Sigma and provides an overview of the key principles underpinning the approach.

We explain exactly what a 'sigma' is and introduce the commonly used process improvement method known as DMAIC – Define, Measure, Analyse, Improve, and Control.

Chapter 1

Defining Lean Six Sigma

In This Chapter

- Turning up trumps for the Toyota Production System
- Finding out the fundamentals of 'lean' and 'six sigma'
- Applying Lean Six Sigma in your organisation

Throughout this book we cover the tools and techniques available to help you achieve real improvement in your organisation. In this chapter we aim to move you down a path of different thinking that gets your improvement taste buds tingling. We look at the main concepts behind lean thinking and six sigma and introduce some of the terminology to help you on your way.

Introducing Lean Thinking

Lean thinking focuses on enhancing value for the customer by improving and smoothing the process flow (see Chapter 11) and eliminating waste (covered in Chapter 9). Since Henry Ford's first production line, lean thinking has evolved through a number of sources, and over many years, but much of the development has been led by Toyota through the Toyota Production System (TPS). Toyota built on Ford's production ideas, moving from high volume, low variety, to high variety, low volume.

Lean is called 'lean' not because things are stripped to the bone. Lean isn't a recipe for your organisation to slash its costs, although it will likely lead to reduced costs and better value for the customer. We trace the concept of 'lean' back to 1987, when John Krafcik (now with Hyundai), was working as a researcher for MIT as part of the International Motor Vehicle Program. Krafcik needed a label for the TPS phenomenon that described what the system did. On a white board he wrote the performance attributes of the Toyota system compared with traditional mass production. TPS:

- Needed less human effort to design products and services.
- Required less investment for a given amount of production capacity.
- Created products with fewer delivered defects.

- Used fewer suppliers.
- Went from concept to launch, order to delivery, and problem to repair in less time and with less human effort.
- Needed less inventory at every process step.
- Caused fewer employee injuries.

Krafcik commented:

> *It needs less of everything to create a given amount of value, so let's call it Lean.*

The lean enterprise was born.

Bringing on the basics of lean

Figure 1-1 shows the Toyota Production System, highlighting various tools and Japanese lean thinking terms that we use throughout this book. In this chapter we provide some brief descriptions to introduce the lean basics and the TPS.

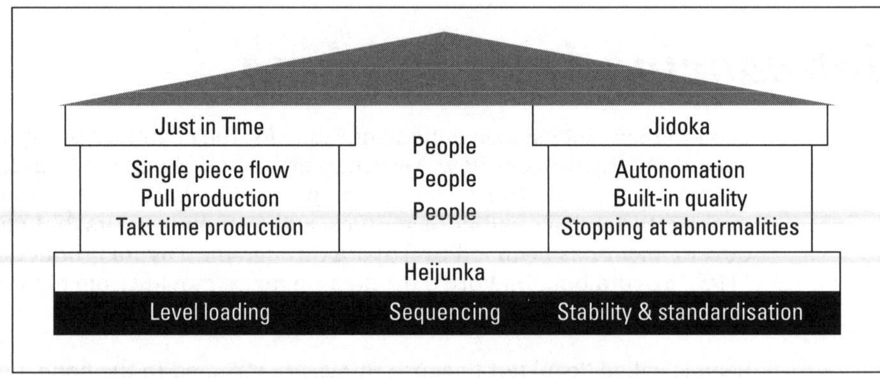

Figure 1-1: The TPS house.

Toyota's Taiichi Ohno describes the TPS approach very effectively:

> *All we are doing is looking at a timeline from the moment the customer gives us an order to the point when we collect the cash. And we are reducing that timeline by removing the non-value-added wastes.*

Picking on people power

Figure 1-1 shows that people are at the heart of TPS. The system focuses on training to develop exceptional people and teams that follow the company's philosophy to gain exceptional results. Consider the following:

- Toyota creates a strong and stable culture where values and beliefs are widely shared and lived out over many years.
- Toyota works constantly to reinforce that culture.
- Toyota involves cross-functional teams to solve problems.
- Toyota keeps teaching individuals how to work together.

Being lean means involving people in the process, equipping them to be able, and feel able, to challenge and improve their processes and the way they work. Never waste the creative potential of people!

Looking at the lingo

You can see from Figure 1-1 that lean thinking involves a certain amount of jargon – some of it Japanese. This section defines the various terms to help you get lean thinking as soon as possible:

- **Heijunka** provides the foundation. It encompasses the idea of smoothing processing and production by considering levelling, sequencing, and standardising:
    - **Levelling** involves smoothing the volume of production in order to reduce variation, that is, the ups and downs and peaks and troughs that can make planning difficult. Amongst other things, levelling seeks to prevent 'end-of-period' peaks, where production is initially slow at the beginning of the month, but then quickens in the last days of the sales or accounting period, for example.
    - **Sequencing** may well involve mixing the types of work processed. So, for example, when setting up new loans in a bank, the type of loan being processed is mixed to better match customer demand, and help ensure applications are actioned in date order. So often, people are driven by internal efficiency targets, whereby they process the 'simple tasks' first to get them out of the way and 'hit their numbers', leaving the more difficult cases to be processed later on. This means tasks are not processed in date order, and a reluctance exists to get down and tackle a pile of difficult cases at the end of the week, making things even worse for the customer and the business.

- **Standardising** is the third strand of Heijunka. It seeks to reduce variation in the way the work is carried out, highlighting the importance of 'standard work', of following a standard process and procedure. It links well to the concept of process management, where the process owner continuously seeks to find and consistently deploy best practice. Remember, however, that you need to standardise your processes before you can improve them. Once they're standardised, you can work on stabilising them, and now that you fully understand how the processes work, you can improve them, creating a 'one best way' of doing them.

  In the spirit of continuous improvement, of course, the 'one best way' of carrying out the process will keep changing, as the people in the process identify better ways of doing the work. You need to ensure the new 'one best way' is implemented and fully deployed.

- **Jidoka** concerns prevention; it links closely with techniques such as Failure Mode Effects Analysis (FMEA) covered in Chapter 10. Jidoka has two main elements, and both seek to prevent work continuing when something goes wrong:

  - **Autonomation** allows machines to operate autonomously, by shutting down if something goes wrong. This concept is also known as automation with human intelligence. The 'no' in auto*no*mation is often underlined to highlight the fact that no defects are allowed to pass to a follow-on process. An early example is from 1902, when Sakichi Toyoda, the founder of the Toyota group, invented an automated loom that stopped whenever a thread broke. A simple example today is a printer stopping processing copy when the ink runs out.

    Without this concept, automation has the potential to allow a large number of defects to be created very quickly, especially if processing is in batches (see 'Single piece flow' later in this section).

  - **Stop at every abnormality** is the second element of Jidoka. The employee can stop an automated or manual line if he or she spots an error. At Toyota every employee is empowered to 'stop the line', perhaps following the identification of a special cause on a control chart (see Chapter 7).

    Forcing everything to stop and immediately focus on a problem can seem painful at first, but doing so is an effective way to quickly get at the root cause of issues. Again, this can be especially important if you're processing in batches.

- **Just in Time (JIT)** provides the other pillar of the TPS house. JIT involves providing the customer with what's needed, at the right time and in the right quantity. The concept applies to both internal and external customers. JIT comprises three main elements:

- **Single piece flow** means each person performs an operation and makes a quick quality check before moving their output to the next person in the following process. If a defect is detected, Jidoka is enacted; the process is stopped, and immediate action is taken to correct the situation, taking countermeasures to prevent reoccurrence. This concept is a real change of thinking that moves us away from processing in batches.

  Traditionally, large batches of individual cases are processed at each step and are passed along the process only after an entire batch has been completed. The delays are increased when the batches travel around the organisation, both in terms of the transport time, and the time they sit waiting in the internal mail system. At any given time, most of the cases in a batch are sitting idle, waiting to be processed. In manufacturing, this is seen as costly excess inventory. What's more, errors can neither be picked up, nor addressed quickly; if they occur, they often occur in volume. And, of course, this also delays identifying the root cause. With single piece flow, we can get to the root cause analysis faster, which helps prevent a common error recurring throughout the process.

- **Pull production** is the second element of JIT. Each process takes what it needs from the preceding process only when it needs it and in the exact quantity. The customer pulls the supply and helps avoid being swamped by items that aren't needed at a particular time.

  Pull production reduces the need for potentially costly storage space. All too often, overproduction in one process, perhaps to meet local efficiency targets, results in problems downstream. This increases work in progress, and creates bottlenecks. Overproduction is one of the 'seven wastes' identified by Ohno and covered in Chapter 9.

- **Takt time** is the third element of JIT, providing an important additional measure. It tells you how quickly to action things, given the volume of customer demand. Takt is German for a precise interval of time, such as a musical meter. It serves as the rhythm or beat of the process – the frequency at which a product or service must be completed in order to meet customer needs. It's a bit like the beat of the drum on the old Roman galleys.

Taking the strain out of constraints

Much of the focus in lean thinking is on understanding and improving the flow of processes and eliminating non-value-added activities. Eli Goldratt's *theory of constraints* (explained more fully in Chapter 11), provides a way to address and tackle bottlenecks that slow the process flow. Goldratt's theory proposes a five-step approach to help improve flow:

1. **Identify the constraint.** Data helps you identify the bottlenecks in your processes, of course, but you should be able to see them fairly easily, too. Look for backlogs and a build-up of work in progress, or take note of where people are waiting for work to come through to them. These are pretty good clues that demand is exceeding capability and you have a bottleneck.
2. **Exploit the constraint.** Look for ways to maximise the processing capability at this point in the process flow. For example, you may minimise downtime for machine maintenance by scheduling maintenance outside of normal hours.
3. **Subordinate the other steps to the constraint.** You need to understand just what the bottleneck is capable of – how much it can produce, and how quickly it can do it. Whatever the answer is, in effect, that's the pace at which the whole process is working. The downstream processes know what to expect and when, and having upstream processes working faster is pointless; their output simply builds up as a backlog at the bottleneck. So, use the bottleneck to dictate the pace at which the upstream activities operate, and to signal to the downstream activities what to expect, even if that means these various activities are not working at capacity.
4. **Elevate the constraint.** Introduce improvements that remove this particular bottleneck, possibly by using a DMAIC (Define, Measure, Analyse, Improve and Control) project (we delve into DMAIC in Chapter 2).
5. **Go back to Step 1 and repeat the process.** After you complete Steps 1–4, a new constraint will exist somewhere else in the process flow, so start the improvement process again.

Considering the customer

The customer, not the organisation, specifies value. Value is what your customer is willing to pay for. To satisfy your customer, your organisation has to provide the right products and services, at the right time, at the right price, and at the right quality. To do this, and to do so consistently, you need to identify and understand how your processes work, improve and smooth the flow, eliminate unnecessary steps in the process, and reduce or prevent waste such as rework.

Imagine the processes involved in your own organisation, beginning with a customer order (market demand) and ending with cash in the bank (invoice or bill paid). Ask yourself the following questions:

- How many steps are involved?
- Do you need all the steps?
- Are you sure?
- How can you reduce the number of steps and the time involved from start to finish?

Perusing the principles of lean thinking

Lean thinking has five key principles:

- Understand the customer and their perception of value.
- Identify and understand the value stream for each process and the waste within it.
- Enable the value to flow.
- Let the customer pull the value through the processes, according to their needs.
- Continuously pursue perfection (continuous improvement).

We've covered these briefly in the preceding pages, but look at them again in more detail in Chapter 2, when we see how they combine with the key principles of six sigma to form *Lean Six Sigma.*

Sussing Six Sigma

Six sigma is a systematic and robust approach to improvement, which focuses on the customer and other key stakeholders. Six sigma calls for a change of thinking. When Jack Welch, former General Electric CEO, introduced six sigma, he said:

> *We are going to shift the paradigm from fixing products to fixing and developing processes, so they produce nothing but perfection or close to it.*

In the 1980s Motorola CEO Bob Galvin struggled to compete with foreign manufacturers. Motorola set a goal of tenfold improvement in five years, with a plan focused on global competitiveness, participative management, quality improvement, and training. Quality engineer Bill Smith coined the name of the improvement measurements: six sigma. All Motorola employees underwent training, and six sigma became the standard for all Motorola business processes.

Considering the core of six sigma

A sigma, or standard deviation, is a measure of variation that reveals the average difference between any one item and the overall average of a larger population of items. Sigma is represented by the lower-case Greek letter σ.

Introducing a simple example

Suppose you want to estimate the height of people in your organisation. Measuring everyone isn't practical, so you take a representative sample of 30 people's heights. You work out the mean average height for the group – as an example, let's say this is 5 foot, 7 inches. You then calculate the difference between each person's height and the mean average height. One sigma, or standard deviation, is the average of those differences. The smaller the number, the less variation there is in the population of things you are measuring. Conversely, the larger the number, the more variation. In our example, imagine the standard deviation is one inch, though it might be any number in theory.

Figure 1-2 shows the likely percentage of the population within plus one and minus one standard deviation from the mean, plus two and minus two standard deviations from the mean, and so on. You can see how your information provides a good picture of the heights of all the people in your organisation. You find that approximately two-thirds of them are between 5 foot 6 inches and 5 foot 8 inches tall, about 95 per cent are in the range 5 foot 5 inches to 5 foot 9 inches, and about 99.73 per cent are between 5 foot 4 inches and 5 foot 10 inches.

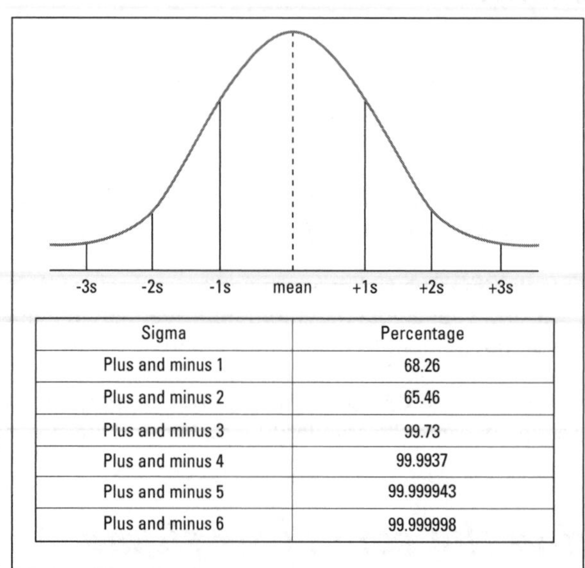

Sigma	Percentage
Plus and minus 1	68.26
Plus and minus 2	65.46
Plus and minus 3	99.73
Plus and minus 4	99.9937
Plus and minus 5	99.999943
Plus and minus 6	99.999998

Figure 1-2: Standard deviation.

In reality, the calculation is a little more involved and uses a rather forbidding formula – as shown in Figure 1-3.

Using n – 1 makes an allowance for the fact that we're looking at a sample and not the whole population. In practice though, when the sample size is over 30, there's little difference between using n or n – 1.

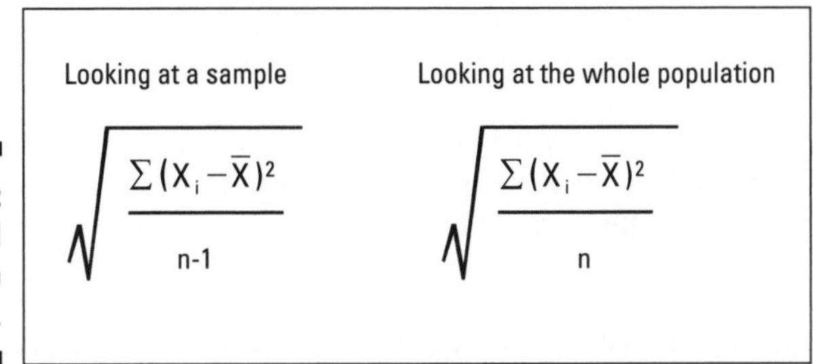

Figure 1-3: Standard deviation formula.

The sigma values are calculated by looking at our performance against the customer requirements – see the next section.

Practising process sigma in the workplace

In the real world you probably don't measure the height of your colleagues. Imagine instead that in your organisation you issue products to customers. You take a representative sample of fulfilled orders and measure the *cycle time* for each order – the time taken from receiving the order to issuing the product (in some organisations this is referred to as *lead time*). Figure 1-4 shows the cycle times for your company's orders.

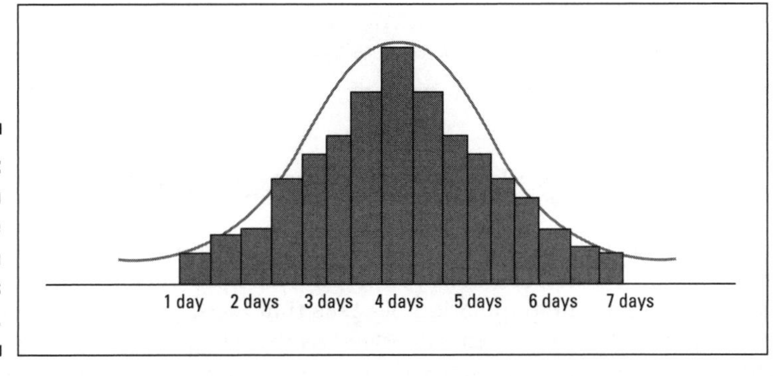

Figure 1-4: Histogram showing the time taken to process orders.

You can see the range of your company's performance. The cycle time varies from as short as one day to as long as seven days.

But the customer expects delivery in five days or less. In Lean Six Sigma speak, a customer requirement is called a CTQ – Critical To Quality. CTQs are referred to in Chapter 2 and described in more detail in Chapter 4, but essentially they express the customers' requirements in a way that is measurable. CTQs are a vital element in Lean Six Sigma and provide the basis of your process measurement set. In our example, the CTQ is five days or less, but the average performance in Figure 1-4 is four days. Remember that this is the average; your customers experience the *whole range* of your performance.

Too many organisations use averages as a convenient way of making their performance sound better than it really is.

In the example provided in Figure 1-4, all the orders that take more than five days are *defects* according to six sigma. Orders that take five days or less meet the CTQ. We show this situation in Figure 1-5. We could express the performance as the percentage of orders processed within five days or we can work out the *process sigma value*. The sigma value is calculated by looking at your performance against the customer requirement, the CTQ, and taking into account the number of 'defects' involved where you fail to meet it (that is, all those cases that took more than five days).

We explain the process sigma calculation in the next section.

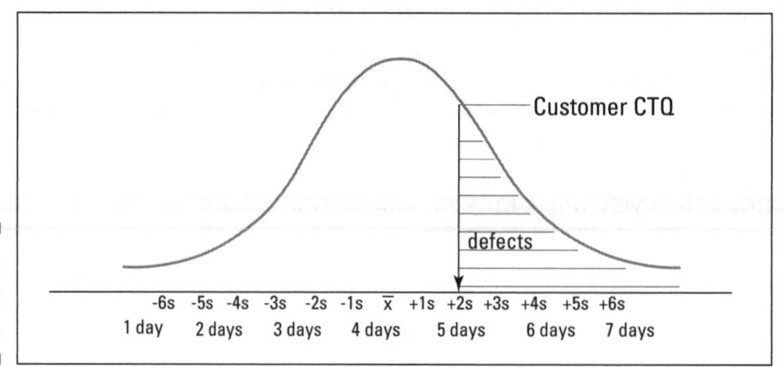

Figure 1-5: Highlighting defects.

Calculating process sigma values

Process sigma values provide a way of comparing performances of different processes, which can help you to prioritise projects. The process sigma value represents the population of cases that meet the CTQs right first time. Sigma values are often expressed as defects per million opportunities (DPMO), rather than per hundred or per thousand, to emphasise the need for world-class performance.

Not all organisations using six sigma calculate process sigma values. Some organisations just use the number of defects or the percentage of orders meeting CTQs to show their performance. Either way, if benchmarking is to be meaningful, the calculations must be made in a consistent manner.

Figure 1-6 includes 'yield' figures – the right first time percentage. You can see that six sigma performance equates to only 3.4 DPMO.

Yield	Sigma	Defects per 1,000,000	Defects per 100,000	Defects per 10,000	Defects per 1,000	Defects per 100
99.99966%	6.0	3.4	0.34	0.034	0.0034	0.00034
99.9995%	5.9	5	0.5	0.05	0.005	0.0005
99.9992%	5.8	8	0.8	0.08	0.008	0.0008
99.9990%	5.7	10	1	0.1	0.01	0.001
99.9980%	5.6	20	2	0.2	0.02	0.002
99.9970%	5.5	.30	3	0.3	0.03	0.003
99.9960%	5.4	40	4	0.4	0.04	0.004
99.9930%	5.3	70	7	0.7	0.07	0.007
99.9900%	5.2	100	10	1.0	0.1	0.01
99.9850%	5.1	150	15	1.5	0.15	0.015
99.9770%	5.0	230	23	2.3	0.23	0.023
99.670%	4.9	330	33	3.3	0.33	0.033
99.9520&	4.8	480	48	4.8	0.48	0.048
99.9320%	4.7	680	68	6.8	0.68	0.068
99.9040%	4.6	960	96	9.6	0.96	0.096
99.8650%	4.5	1,350	135	13.5	1.35	0.135
99.8140%	4.4	1,860	186	18.6	1.86	0.186
99.7450%	4.3	2,550	255	25.5	2.55	0.255
99.6540%	4.2	3,460	346	34.6	3.46	0.346
99.5340%	4.1	4,660	466	46.6	4.66	0.466
99.3790%	4.0	6,210	621	62.1	6.21	0.621
99.1810%	3.9	8,190	819	81.9	8.19	0.819
98.930%	3.8	10,700	1,070	107	10.7	1.07
98.610%	3.7	13.900	1.390	139	13.9	1.39
98.220%	3.6	17,800	1,780	178	17.8	1.78
97.730%	3.5	22.700	2,270	227	22.7	2.27
97.130%	3.4	28,700	2,870	287	28.7	2.87
96.410%	3.3	35,900	3,590	359	35.9	3.59
95.540%	3.2	44,600	4,460	446	44.6	4.46
94.520%	3.1	54,800	5,480	548	54.8	5.48
93.320%	3.0	66,800	6,680	668	66.8	6.68
91.920%	2.9	80,800	8,080	808	80.8	8.08
90.320%	2.8	96,800	9.680	968	96.8	9.68
88.50%	2.7	115,000	11,500	1,150	115	11.5
86.50%	2.6	135,000	13,500	1,350	135	13.5
84.20%	2.5	158,000	15,800	1,580	158	15.8
81.60%	2.4	184,000	18,400	1,840	184	18.4
78.80%	2.3	212,000	21,200	2,120	212	21.2
75.80%	2.2	242,000	24,200	2,420	242	24.2
72.60%	2.1	274,000	27,400	2,740	274	27.4
69.20%	2.0	308,000	30,800	3,080	308	30.8
65.60%	1.9	344,000	34,400	3,440	344	34.4
61.80%	1.8	382,000	38,200	3,820	382	38.2
58.00%	1.7	420,000	42,000	4,200	420	42
54.00%	1.6	460,000	46,000	4,600	460	46
50%	1.5	500,000	50,000	5,000	500	50
46%	1.4	540,000	54,000	5,400	540	54
43%	1.3	570,000	57,000	5,700	570	57
39%	1.2	610,000	61,000	6,100	610	61
35%	1.1	650,000	65,000	6,500	650	65
31%	1.0	690,000	69,000	6,900	690	69
28%	0.9	720,000	72,000	7,200	720	72
25%	0.8	750,000	75,000	7,500	750	75
22%	0.7	780,000	78,000	7,800	780	78
19%	0.6	810,000	81,000	8,100	810	81
16%	0.5	840,000	84,000	8,400	840	84
14%	0.4	860,000	86,000	8,600	860	86
12%	0.3	880,000	88,000	8,800	880	88
10%	0.2	900,000	90,000	9,000	900	90
8%	0.1	920,000	92,000	9,200	920	92

Figure 1-6: Abridged sigma conversion table.

Recognising that you're looking at 'first pass' performance is important. If you make an error but correct it before the order goes to the customer, you still count the defect because the rework activity costs you time and effort. And remember that you're looking at defects. Your customer may have several CTQs relating to an order – for example, speed and accuracy – thus there may be more than one defect in the transaction.

In calculating sigma values for your processes, you need to understand the following key terms:

- **Unit:**The item produced or processed.
- **Defect:** Any event that does not meet the specification of a CTQ.
- **Defect opportunity:** Any event that provides a chance of not meeting a customer CTQ. The number of defect opportunities will equal the number of CTQs.
- **Defective:** A unit with one or more defects.

In manufacturing processes you may find that the number of defect opportunities is determined differently, taking full account of all the different defects that can occur within a part. The key is to calculate the sigma values in a consistent way.

You can work out your sigma performance against the CTQs as shown in Figure 1-7. We have a sample of 500 processed units. The customer has three CTQs, so we have three defect opportunities. The CTQs are related to speed, accuracy, and completeness. We find 57 defects.

- Number of units processed — N=500
- Total number of defects made (include defects made and later fixed) — D=57
- Number of defect opportunities per unit (equate to CTQs) — O=3
- Calculate # defects per million opportunities

$$\text{DPMO} = 1{,}000{,}000 \times \frac{D}{(N \times O)}$$

$$= 1{,}000{,}000 \times \frac{57}{(500) \times (3)}$$

$$= 38000$$

- Look up process sigma in sigma conversation table (see Figure 1-6) — Sigma = 3.3

Figure 1-7: Calculating process sigma values.

A difference exists between sigma and standard deviation (see the 'Introducing a simple example' section earlier in this chapter for how to work out standard deviations). This results from Motorola adjusting the tables to reflect the variation being experienced in their processes. This adjustment is referred to as a 1.5 sigma shift, reflecting the extent of the adjustment. Although this related to *their* processes, everyone adopting six sigma also adopted the adjusted sigma scale. Incidentally, without this adjustment, six sigma would equate to 0.002 DPMO as opposed to 3.4 DPMO – so, even harder to achieve.

When we talk about six sigma performance before the adjustment, we're talking about plus and minus six standard deviations, which embrace 99.999998 per cent of the data. And we are talking about the percentage of cases that are right first time in terms of meeting the requirements of the customer. Taking account of the adjustment, we're still looking at a truly demanding standard, with 99.999666 per cent of cases right first time.

Meeting the major points of six sigma

The five key principles of six sigma are:

- **Understand the CTQs of your customers and stakeholders.** To deliver the best customer experience, you need to know what your customer wants – their requirements and expectations. You need to listen to and understand the *voice of the customer* (VOC) – we talk about the customer's voice in Chapter 4.
- **Understand your organisation's processes and ensure they reflect your customers' CTQs.** You need to know how your processes work and what they're trying to achieve. There should be a clear objective for each process focused on the customer requirements, the CTQs.
- **Manage by fact and reduce variation.** Measurement and management by fact enables more effective decision making. By understanding variation, you can work out when and when not to take action.
- **Involve and equip the people in the process.** To be truly effective you need to equip the people in your organisation to be able, and to feel able, to challenge and improve their processes and the way they work.
- **Undertake improvement activity in a systematic way.** Working systematically helps you avoid jumping to conclusions and solutions. Six sigma uses a system called DMAIC (Define, Measure, Analyse, Improve and Control) to improve existing processes. We cover DMAIC in Chapter 2.

A natural synergy exists between lean and six sigma – your organisation needs both. Many people think of lean as focusing on improving the efficiency of processes, and six sigma as concentrating on their effectiveness. The reality is that both approaches tackle efficiency and effectiveness.

Chapter 2

Understanding the Principles of Lean Six Sigma

In This Chapter

- Merging lean and six sigma to make Lean Six Sigma
- Doing the DMAIC to make things better
- Reviewing what you do in order to do it better

In this chapter we look at the synergy produced by combining the approaches of lean and six sigma to form Lean Six Sigma. The merged approach provides a comprehensive set of principles, and supporting tools and techniques to enable genuine improvements in both efficiency and effectiveness for organisations.

Considering the Key Principles of Lean Six Sigma

Lean Six Sigma takes the features of lean and of six sigma and integrates them to form a magnificent seven set of principles. The principles of each approach aren't dissimilar (check out Chapter 1 to read more about the individual components), and the merged set produces no surprises. The seven principles of Lean Six Sigma are:

- **Focus on the customer.** The customer's CTQs describe elements of your service or offering they consider Critical To Quality (see Chapter 1 for more on these). Written in a way that ensures they are measurable, the CTQs provide the basis for determining the process measures you need to help you understand how well you perform against these critical requirements. Focusing on the customer and the concept of value-add is important because typically only 10–15 per cent of process steps add value and often represent only 1 per cent of the total process time. These figures may be surprising, but they should grab your attention and help you realise the potential waste that is happening in your own organisation. As you improve your performance in meeting the CTQs, you're also likely to win and retain further business and increase your market share. The concept of value-add is covered in Chapter 9 and Chapters 3 to 5 consider the customer in more detail.

- **Identify and understand how the work gets done.** The *value stream* describes all of the steps in your process. By drawing a map of the value stream, you can highlight the non-value-added steps and areas of waste and ensure the process focuses on meeting the CTQs and adding value. To undertake this process properly, you must 'go to the Gemba'. The Japanese word Gemba means the place where the work gets done – where the action is – which is where management begins. Process stapling (which we introduce in Chapter 5) involves you spending time in the workplace to see how the work really gets done, not how you think it gets done or how you'd like it to be done. You see the real process being carried out and collect data on what's happening. Process stapling helps you analyse the problems that you want to tackle and determines a more effective solution for your day-to-day activities.

The value stream reveals all of the actions, both value-creating and non-value-creating, that take your product or service concept to launch and your customer order through the supply chain to delivery. These value-creating and non-value-creating actions include those to process information from the customer and those to transform the product on its way to the customer. Chapter 5 covers the value stream.

- **Manage, improve, and smooth the process flow.** The concept of managing, improving, and smoothing the process flow provides an example of different thinking. If possible use single piece flow, moving away from batches, or at least reducing batch sizes. Either way, identify the non-value-added steps in the process and try to remove them – certainly look to ensure they do not delay value-adding steps. The concept of pull, not push (see Chapter 1), links to our understanding the process and improving flow. And it can be an essential element in avoiding bottlenecks. Overproduction or pushing things through too early is a waste.

- **Remove non-value-add steps and waste.** Doing so is another vital element in improving flow and performance, generally. The Japanese refer to waste as Muda; they describe two broad types and seven categories of waste. Of course, if you can prevent waste in the first place, then so much the better (see Chapters 9 and 10 on how to do this).

- **Manage by fact and reduce variation.** Managing by fact, using accurate data, helps you avoid jumping to conclusions and solutions. You need the facts! And that means measuring the right things in the right way. Data collection is a process and needs to be managed accordingly. Using Control Charts (Chapter 7 has more on these) enables you to interpret the data correctly and understand the process variation. You then know when to take action and when not to.

- **Involve and equip the people in the process.** You need to involve the people in the process, equipping them to both feel and be able to challenge and improve their processes and the way they work. It's what has to be done if organisations are to be truly effective, but like so many of the Lean Six Sigma principles, it requires different thinking if it's to happen. (See Chapter12 for more on understanding the 'people issues'.)
- **Undertake improvement activity in a systematic way.** DMAIC comes into play here: Define, Measure, Analyse, Improve, and Control. One of the criticisms sometimes aimed at 'stand-alone' lean is that improvement action tends not to be taken in a systematic and standard way. In six sigma, DMAIC is used to improve existing processes, but the framework is equally applicable to lean and, of course, Lean Six Sigma.

Less is usually more. Tackle problems in bite-size chunks and never jump to conclusions or solutions. (Chapter 14 highlights the danger of jumping to conclusions.)

Improving Existing Processes: Introducing DMAIC

DMAIC (Define, Measure, Analyse, Improve, and Control) provides the framework to improve existing processes in a systematic way. DMAIC projects begin with the identification of a problem, and in the Define phase you describe what you think needs improving. Without data this might be based on your best guess of things, so in the Measure phase you use facts and data to understand how your processes work and perform so that you can describe the problem more effectively.

Now you can Analyse the situation by using facts and data to determine the root cause(s) of the problem that is inhibiting your performance. With the root cause identified you can now move to the Improve phase and identify, select, and implement the most appropriate solution(s), validating your approach with data.

The Control phase is especially important. You can check that your customers feel the difference in your performance and you'll need to use data to help you hold the gains. After all your hard work, you don't want the problem you've solved to recur. With the right ongoing measures in place, you should also be able to prompt new opportunities (see Chapter 8 for more on getting the right balance of measures).The following sections provide a little more detail about the five DMAIC phases. Figure 2-1 shows how the phases link together.

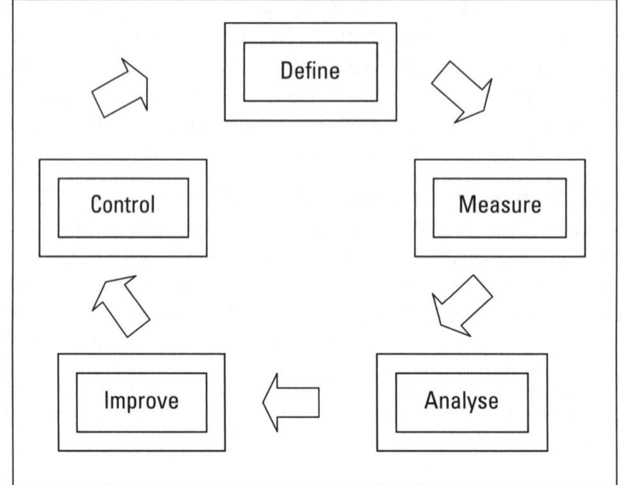

Figure 2-1: The five phases of DMAIC.

Defining your project

When you start an improvement project, ensuring that you and your team understand why you're undertaking the project and what you want to achieve is an essential ingredient for success. With a DMAIC project, you start with a problem that needs to be solved. Before you can solve the problem, you need to define it – not always as straightforward a process as you might think. One of the key outputs from the Define phase is a completed *improvement charter.*

The improvement charter is an agreed document defining the purpose and goals for an improvement team. It can help address some of the elements that typically go wrong in projects by providing a helpful framework to gain commitment and understanding from the team. Keep your charter simple and try to contain the document to one or two sides of A4 in line with the example shown in Figure 2-2.

The improvement charter contains the following key elements:

- **A high-level business case** providing an explanation of why undertaking the project is important.
- **A problem statement** defining the issue to be resolved.
- **A goal statement** describing the objective of the project.
- **The project scope** defining the parameters and identifying any constraints.
- **The CTQs** specifying the problem from the customers' perspective. Unless you already have the CTQs, these might not be known until the measure phase.

- **The project scope** defining the parameters and identifying any constraints.
- **The CTQs** specifying the problem from the customers' perspective. Unless you already have the CTQs, these might not be known until the measure phase.
- **Roles** identifying the people involved in and around the project, expectations of them, and their responsibilities. The improvement charter forms a contract between the members of the improvement team.
- **Milestones** summarising the key steps and provisional dates for achieving the goal.

Improvement Charter

Project title:	Date commenced:
Why *High level business case describing why this project is important and how it links to our business plans*	
What *The problem and goal statements, the scope, and the CTQ and defect definitions for the relevant customers and processes*	
Problem statement	Goal statement
In frame	Out of frame
CTQs	Defect definition
Who *The process owner, Champion, team leader, and team members. Who are they and what are their roles, responsibilities, and time commitments? What involvement is expected of the Champion? How often should they meet?*	

Name	Roles & Responsibilites	Time commitment

When *High level timeframes for the phases. This could be mapped to the eight steps.*

	Date	Date	Date	Date	Date	Date
Define						
Measure						
Analyse						
Improve						
Control						

Figure 2-2: A sample improvement charter.

The improvement charter needs to be seen as a 'living document' and be updated through the various DMAIC phases, especially as your understanding of the problem you're tackling becomes clearer.

Depending on the nature of your project, you may also need to use some other tools, such as *affinity* and *interrelationship diagrams* which we describe in a moment (see Figures 2-3 and 2-4). If your project is large and potentially complex, an affinity diagram prepares you for success. It can also aid you in developing your improvement charter. Affinity and interrelationship diagrams provide definition for your project and help the team really understand what's involved. These tools should be used together. The affinity diagram can be the first step in a large project (we like to think of it as 'step zero') and it helps the team develop their thoughts on the issues involved. By the time they've created the interrelationship diagram, the team will have a detailed understanding of what they need to do, the drivers of success, and the many and varied interrelationships involved, and they will feel they own the output from the exercise.

Figure 2-3 shows the steps in the creation of an affinity diagram. The process works best if you use sticky notes and silently brainstorm ideas on an agreed *issue statement*; for example, 'what issues are involved in introducing Lean Six Sigma into our organisation?' Follow these rules:

- Use one idea per sticky note.
- Write statements rather than questions.
- Write clearly.
- Don't write in upper case (reading lower-case words is easier).
- Avoid one-word statements (your colleagues won't know what you mean).
- Include a noun and verb in each statement.
- Don't write an essay.

Once everyone has finished writing their sticky notes, maintain the silence and place them on the wall, as shown in the first part of Figure 2-3. Move the notes into appropriate themes or clusters (see the middle figure). Finally, give each theme or cluster a title describing its content (see the final figure). Ensure that each title provides enough description; doing so is helpful for when you move into the interrelationship diagram, shown in Figure 2-4.

An *interrelationship diagram* identifies the key causal factors or drivers for your programme or project, by enabling you to understand the relationships between the themes or clusters. In looking at the different pairs of clusters you're looking to see if a cause and effect relationship exists.

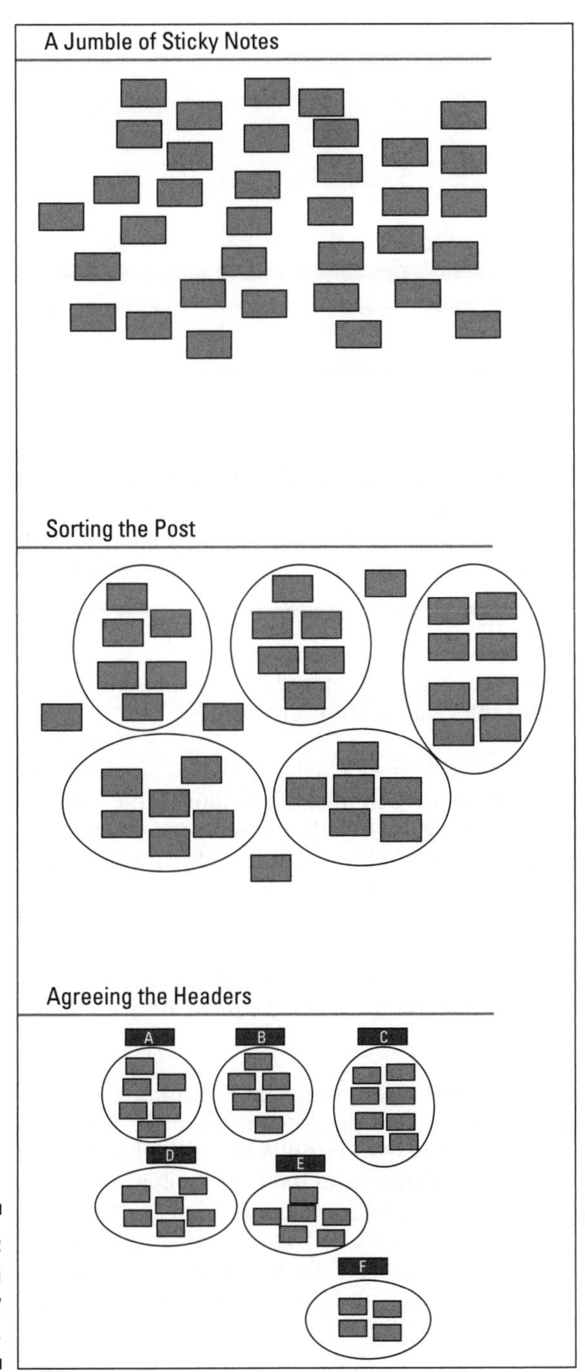

Figure 2-3: Creating an affinity diagram.

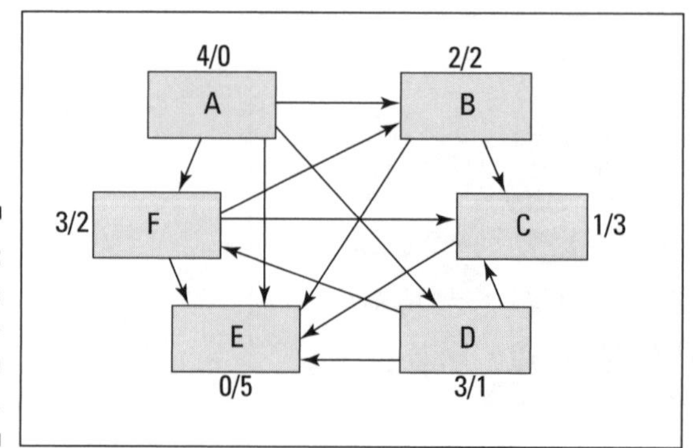

Figure 2-4: A sample interrelationship diagram.

In Figure 2-4, the headers for the themes or clusters have been put into a clock face on a flip chart and you now work your way round looking at the relationship between each pair. As you do so, you need to consider whether a relationship exists or not, and, where it does, determine which has a greater effect on the other – for example, 'must this happen before this does?'

If a relationship does exist between two clusters, connect them with a line. Importantly, either there is a relationship or there isn't, so don't use dotted lines – you might start with a pencil, though! It's sensible to agree some ground rules to ensure the relationships are tangible.

After you've determined the 'causal' cluster, draw an arrow into the 'effect' cluster. Some discussion is likely to take place about which way the arrow faces, but it has to go one way or the other – two-headed arrows are not allowed! In Figure 2-4, you can see that 'A' drives 'B', but that 'B' is the driver of 'C'. The numbers represent arrows out over arrows in.

The finished diagram can be presented as shown in Figure 2-5, and you can clearly see the key driver is 'A', whereas 'E' is probably the outcome of the project. You need to particularly focus on 'A' to ensure your project or programme is successful.

Throughout your project, developing a storyboard summary of the key decisions and outputs helps you review progress and share what you've learnt. A storyboard builds up as you work your way through your project by capturing the key outputs and findings from the DMAIC phases. A storyboard would include, for example, your improvement charter and process map (see Chapter 5). The storyboard also helps your communication activities. Developing and reviewing a communications plan is an essential activity. You really need to keep your team and the people affected by your project informed about the progress you're making in solving the problem you're tackling. Communication begins on day one of your project.

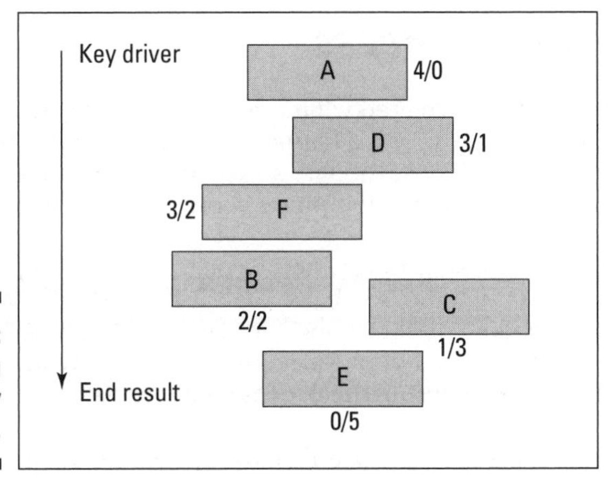

Figure 2-5: Identifying the key drivers.

Measuring how the work is done

After you've defined the problem, at least based on your current understanding, you need to clarify how, and how well, the work gets done. To understand the current situation of your process, knowing what it looks like and how it's performing is important. You need to know what's meant to happen, and why. Understanding how your process links to your customer and their CTQs is also helpful. What does the bigger picture of the process look like?

Knowing the current performance of your process is essential – this knowledge becomes your baseline – but knowing what's happened in the past is also useful. Measure what's important to the customer, and remember also to measure what the customer sees. Gathering this information can help focus your improvement efforts and prevent you going off in the wrong direction. Using *control charts* (see Chapter 7) can help you make better sense of the data, as they provide a visual picture that demonstrates performance and shows you the variation within the process. Importantly, control charts help you know when to take action and when not to by enabling you to identify the key signals so often hidden when data is presented as a page of numbers.

Six sigma projects can take longer than you might like because the right data isn't in place in the day-to-day operation. So often, organisations have data coming out of their ears – but not the *right* data. You need to develop the right measures and start collecting the data you do need – which takes time.

Use the CTQs as the basis for getting the right process measures in place. Understanding how well you meet the CTQs is an essential piece of management information. Chapter 8 provides more detail on getting the right measures.

Analysing your process

In the Measure phase, you discovered what's really happening in your process. Now you need to identify why it's happening, and determine the root cause. You need to manage by fact, though, so you must verify and validate your ideas about possible suspects. Jumping to conclusions is all too easy.

Carrying out the Analyse phase properly helps you in determining the solution when you get to the Improve phase. Clearly, the extent of analysis required varies depending on the scope and nature of the problem you're tackling, and, indeed, what your Measure activities have identified. Essentially, though, you're analysing the process and the data.

Checking the possible causes of your problem using concrete data to verify your ideas is crucial. In checking the vital few, you may find the 'usual suspects' aren't guilty at all! Identifying and removing the root causes of a problem prevents it happening again.

Improving your process

Most people want to start at this point. Now you've identified the root cause of the problem, you can begin to generate improvement ideas to help solve it. Improve, however, involves three distinct phases:

1. **Generate ideas about possible solutions.** The solution may be evident from the work done in the previous two steps. Make sure that your proposed solutions address the problem and its cause.
2. **Select the most appropriate solution.** Take account of the results from any testing or piloting, and the criteria you've identified as important, such as customer priorities, cost, speed, and ease of implementation. Ensure your solution addresses the problem and that customers will see a difference if you adopt it.
3. **Plan and test the solution.** This step seeks to ensure the smooth implementation of your chosen solution. The main focus, though, is on prevention – causing something not to happen. Carrying out a more detailed pilot is likely to be helpful.

Coming up with a control plan

After all your hard work, you need to implement the solution in a way that ensures you make the gain you expected and hold onto it! If you're to continue your efforts in reducing variation and cutting out waste, the changes being made to the process need to be consistently deployed and followed.

If the improvement team is handing over the 'new' process to the process team, the handover process needs to ensure that everyone understands who's responsible for what and when. Misunderstandings are all too easy and a clear cut-off point must exist signalling the end of the improvement team's role. A control plan needs to be developed to ensure that the gain is secured and the new process effectively deployed – see Figure 2-6.

Process	Performance	Action
Deployment Flowchart	Checks and Measures	Corrective Actions
	Plot time on each step; should be two hours or less; check for special causes	If time exceeds two hours, alert team leader and organize investigation
	Count errors	If more than one per oder, stop process, contact team leader and investigate

Figure 2-6: A control plan helps to make the handover process run smoothly.

The *control plan* helps to ensure that the process is carried out consistently. It also identifies key points in the process where measurement data is needed, additionally highlighting what action is required depending on the results. Ensuring you have the right ongoing measures in place is extremely important.

Your various DMAIC projects help the organisation move towards a situation where processes are completely understood, and are focused on meeting customer CTQs involving the minimum possible non-value-added activity.

Reviewing Your DMAIC Phases

Informal reviews of the progress on your improvement project on a weekly or even daily basis may be very sensible, but, as a minimum, you should conduct a formal tollgate review at the end of each DMAIC phase. A *tollgate review* checks that you have completed the current phase properly and reviews the team's various outputs from it. The improvement team leader and the management sponsor or champion of the improvement activity should conduct this review. In effect you're passing through a tollgate.

Before moving from one phase to another, stepping back, assessing progress, and asking some key questions is crucial; for example:

- How are things going; for example, is the team working well together?
- Are we on course?
- What have we discovered?
- What went well? Why?
- What conclusions can we draw?

The tollgates also provide an opportunity to update your improvement charter and storyboard. Doing so pulls together some of the key elements of your project; for example, a picture of the process and a control chart showing performance. The tollgate also enables you to take stock of the benefits accruing and the financial details; for example, reductions in errors, improvements in processing time, and customer satisfaction. In determining the benefits and financial details, ensure you record the assumptions behind your estimates or calculations, as you may need to explain these to others in the organisation.

At the end of the Analyse phase, the review is of particular importance. It provides an opportunity to review the scope of your project, that is, how much improvement you're seeking to achieve from it.

Before the project began, you may well have best-guessed a business case that justifies starting the work. By the end of this phase, you should be able to *Quantify the Opportunity* – to really understand the extent of non-value-add activities and waste, and the potential for improvement. On completion of the Measure phase, you're able to understand the current situation and level of performance. Following the Analyse phase, your level of understanding will have increased significantly and you understand the root cause of the problem:

- You know why performance is at the level it is.
- You understand the costs involved in the process, both overall and at the individual step level.
- You have identified the waste and the non-value-add steps, including the extent of rework and understood their impact on your ability to meet the CTQs.

In Quantifying the Opportunity, you first need to calculate the saving if all this waste and non-value-added work were eliminated, making sure you document your assumptions. You may feel the opportunity is too small to bother about, or so large it justifies either widening the scope of the project or developing a phased approach, by breaking the task into several smaller projects, for example. Either way, review and agree your project goals now, sensibly estimating what is possible for your project.

The benefits are reviewed again closely following your completion of the Improve phase. You're looking to confirm the deliverables from the project, and secure authority for the solution to be fully implemented. As with Quantify the Opportunity, the post-improve review also provides an opportunity to look at the project more generally, and key questions include:

- Are we on course?
- What have we discovered? And forgotten?
- What went well? Why?
- Can we apply the solution elsewhere?
- What conclusions can we draw?

Confirming the benefits you expect to achieve is the main focus of this second benefits review, for example in reduced re-work or improved processing speeds. In completing the phase, you should feel confident that the chosen solution addresses the root cause of the identified problem, and ensure you meet the project goals. Management by fact is a key principle of Lean Six Sigma so you should have appropriate measurement data and feel confident that your solution will deliver.

Quite a range of differing benefits may occur, including:

- Reduced errors and waste.
- Faster cycle time.
- Improved customer satisfaction.
- Reduced cost.

In assessing how well these benefits match the project objectives, bear in mind that quantifying the softer benefits of enhanced customer satisfaction may be difficult. And in projecting when the benefits are likely to emerge, don't lose sight of the fact that a time gap will probably exist between the cause and effect, especially where customer satisfaction feedback and information is concerned.

As well as looking at the benefits, this review also confirms any costs associated with the solution and its implementation. The piloting or testing activity carried out in the Improve phase (see 'Improving your process' earlier in this chapter) should have helped you pull this information together, provided you treated it as though it were a full-scale implementation. Internal guidelines will probably be available to help you assess and present the benefits and costs, but ensure you've documented the assumptions behind your benefits assessment.

A third and final benefit review follows the Control phase, enabling you to confirm the actual costs and benefits and whether any unexpected debits or credits have occurred. And you should know the answers to these questions:

- Do our customers feel there has been an improvement? How do we know?
- Can we take any of the ideas or 'best practices' and apply them elsewhere in the business?

This review is the formal post-implementation phase involving the project sponsor or champion. In some organisations you may find a wider team of managers forming a 'project board' or 'steering committee', which help provide overall guidance for improvement teams and help prevent duplication of effort with different teams tackling the same or similar problems. This review is likely to involve your team presenting their storyboard, as described in the 'Defining your project' section, earlier in this chapter.

Taking time for these reviews and tollgates is an important element in developing a culture that manages by fact. Maintaining an up-to-date storyboard as you work your way through the DMAIC phases helps you prepare for the reviews and share discoveries. The storyboard is created by the team and should present the important elements of its work – the key outputs from the DMAIC process.

Taking a Pragmatic Approach

Six sigma and DMAIC have been criticised for being too complex, and for projects taking too long. Be pragmatic. Projects need to take as long as is appropriate and often only a few simple tools and techniques are needed to secure quick and successful improvements.

Some say that lean doesn't always ensure a systematic and controlled approach to achieving and holding on to improvement gains – this is where the Control phase of DMAIC is so important. For relatively straightforward problems, *rapid improvement events* can be utilised. These can be run in one-week sessions – sometimes known as *Kaizen blitz events*. Kaizen is the Japanese word for continuous improvement; it means 'change for the better'. The implementation of the improvement may take a further month or so, and some pre-event planning and data collection is necessary.

These events bring together the powerful concepts of Kaizen to involve people in continuously seeking to improve performance within the framework of DMAIC. That improvement comes from focusing on how the work gets done and how well.

Wax on, wax off: Lean Six Sigma and martial arts

The different levels of training in Lean Six Sigma are often referred to in terms of the coloured belts acquired in martial arts. Think about the qualities of a martial arts Black Belt: highly trained, experienced, disciplined, decisive, controlled and responsive, and you can see how well this translates into the world of making change happen in organisations. Thankfully, no bricks need to be broken in half by hand . . .

- Some organisations develop a pool of **Yellow Belts** who typically receive 2 days of practical training to a basic level on the most commonly used tools in Lean Six Sigma projects. They work either as project team members or carry out mini-projects themselves in their local work environment under the guidance of a Black Belt.
- **Green Belts** are trained on the basic tools and lead fairly straightforward projects. Foundation Green Belt level (4 to 6 days training) covers lean tools, process mapping techniques, measurement as well as a firm grounding in the DMAIC methodology and the basic set of statistical tools. Advanced Green Belts (an additional 6 days' training) receive further instruction on more analytical statistical tools and start to use statistical software. Green Belts typically devote the equivalent of about a day a week (20 per cent of time) to Lean Six Sigma projects, mentored by a Black Belt.
- An expert level Lean Six Sigma practitioner is trained to **Black Belt** level which means attending several modules of training over a period of months. Most Black Belt courses involve around 20 days of full-time training as well as working on projects in practice, under the guidance of a Master Black Belt.

  The role of the Black Belt is to lead complex projects and provide expert help with the tools and techniques to the project teams.

  Black Belts are often from different operational functions across the company, coming into the Black Belt team from customer service, finance, marketing or HR for example. They stay in the Black Belt role for a term of 2 to 3 years, after which they return into operations. In effect they become internal consultants working on improving the way the organisation works; changing the organisational systems and processes for the better.
- The **Master Black Belt (MBB)** receives the highest level of training and becomes a full-time professional Lean Six Sigma expert. An experienced MBB is likely to want to take on this role as a long-term career path, becoming a trainer, coach or deployment advisor, and working with senior executives to ensure the overall Lean Six Sigma programme is aligned to the strategic direction of the business. MBBs tend to move around from one major business to another after typically 3 or 4 years in one organisation.

Certification processes are operated in many organisations to ensure a set standard is reached through exams and project assessments. Certification processes are established in many countries, such as the British Quality Foundation and the American Society of Quality. Many large corporate businesses set up their own internal certification processes with recognition given at high profile company events to newly graduated 'belts'.

Rapid improvement events can also be run as a series of half- or one-day workshops, over a period of five or six weeks. They follow the DMAIC framework, and particular emphasis is placed on the Define and Control phases. So, for example, the first workshop focuses on getting a clear definition of the problem to be tackled (Define), and so on. The aim is to tackle a closely scoped bite-sized problem using the expertise of the people actually involved in that process to solve it. They'll need someone with Lean Six Sigma experience, as a facilitator as they may need help using some of the tools and techniques required, for example, Value Stream Maps (see Chapter 5). So often, the people doing the job know what's needed to put things right. You may well find that the solution is already known by the team, but historically they haven't been listened to. Implementation of the solution can thus be actioned quickly, much of it during the actual event.

Rapid improvement events provide them with the opportunity to use their skills and knowledge.

In terms of time, the short duration of rapid improvement events compares to perhaps four months part time in a traditional DMAIC project, though the actual team hours may be similar.

As with a traditional DMAIC project, the Control phase is vital to ensure the improvement gain is maintained. Typical lean improvement activities often neglect it.

Part II

Working with Lean Six Sigma

In this part . . .

Throughout the book, we keep encouraging you to ask yourself how and why things are done. What's the purpose of your products and services and the processes that support them?

Ideally, you do certain things to meet the requirements of your customers, but you need to know who they are, or who they might be.

This part focuses on identifying your various and often quite different customers, seeing how you can determine their requirements, and showing how to use this information to form the basis of the measurement set for your processes. In doing so, you need to take a brief look at some process basics, too. By drawing a process map you can see what the process really looks like, and understand who does what, when, where, and why.

In a nutshell, you're developing a picture of your customers and the processes that seek to meet their requirements.

Chapter 3

Identifying Your Customers

In This Chapter

- Finding out the fundamentals of processes
- Engaging with internal and external customers
- Creating a map of your processes

All organisations have a whole range of different customers – internal and external, large and small. Each organisation's processes should be designed and managed in a way that meets their customers' various needs. In this chapter we help you understand some process fundamentals necessary to focusing on your customers and their requirements.

Understanding the Process Basics

A process is a series of steps and actions that produce an output in the form of a product or a service.

All work is a process, and a process is a blend of PEMME:

- ***P*****eople:** those working in or around the process. Do you have the right number in the right place, at the right time, and possessing the right skills for the job?
- ***E*****quipment:** the various items needed for the work. Items can be as simple as a stapling machine or as complicated as a lathe used in manufacturing. Consider whether you have the right equipment, located in an appropriate and convenient place, and being properly maintained and serviced.
- ***M*****ethod:** how the work needs to be actioned – the process steps, procedures, and activities involved.
- ***M*****aterials:** the things necessary to do the work, for example the raw materials needed to make a product.

- **Environment:** covers the working area, where perhaps a room or surface needs to be dust-free, or where room temperature must be within defined parameters.

Focusing on PEMME helps you think differently when considering what a process actually is. (All the elements of PEMME also combine to influence the results from your processes in relation to variation – as covered in Chapter 7.)

Pinpointing the elements of a process

The concept of processes and PEMME applies to everything you do, from getting up in the morning, to making a cup of tea, to paying a bill. All of these activities can be broken down into a series of steps. The process model shown in Figure 3-1 has PEMME at its heart (the 'process'), but it also builds on PEMME and helps you think about the wider requirements of the process. To meet your customers' requirements (the CTQs – Critical to Quality – see Chapters 2 and 4), the process elements must be addressed.

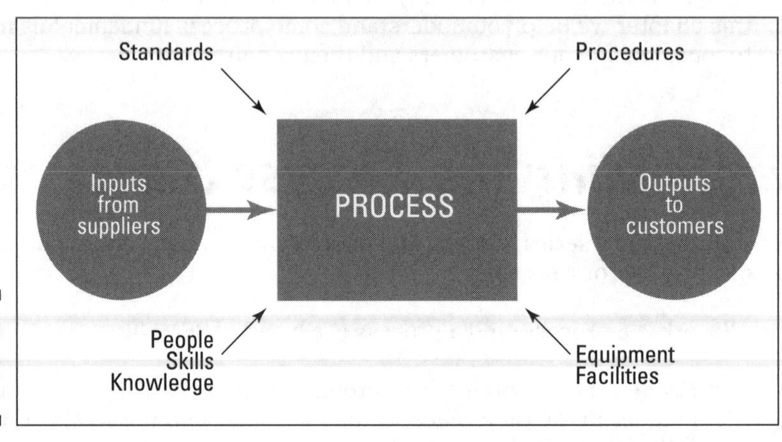

Figure 3-1: Using a process model.

Ensuring that the CTQs are understood and agreed on is the first requirement of a process. More often than not, a lack of quality or re-work is the direct result of not defining the customer's requirements properly. Even on apparently simple things, a little extra time spent in translating the voice of the customer and clarifying requirements can help save time and potential upset later on. Once the customer's requirements have been agreed, determining your own requirements from suppliers is the next step. Now you're the customer, so spend time with your suppliers to ensure your needs are properly understood and agreed.

Make sure you have the right number of people working in the process, and that they have the necessary knowledge and skills. If they don't, appropriate training needs to be delivered.

You need procedures, too. These should describe precisely how the work gets done, the method, and must be developed, agreed, appropriately documented, and kept up to date. Importantly, they should be simple to follow and understand. Clear language and diagrams help ensure they're used to good effect.

Properly describing relevant standards is also sensible. They may well form part of the method, applying, for example, to regulatory requirements or service level agreements that need to be followed. Like procedures the standards should be documented in an easily accessible manner. Similarly, if budget constraints or authority limits on certain actions apply, management must ensure the people in the process know the details.

Equipment and facilities are needed to operate the process. These must be appropriate from day one, located in the right place and correctly maintained thereafter. Also ensure the environment is appropriate for the activity. The facilities link to the environment element of PEMME and could include having the right workspace, for example.

Identifying internal and external customers

All of your processes are likely to involve other people. Some of them probably work in your organisation – your internal customers and suppliers. They're the people involved in the different steps of your process – they may be members of your team or department, but could also be in other departments or functions. An internal supplier provides you with the inputs you need to start your work; for example, information, perhaps a schedule of available products, or an approved order. You're their internal customer.

Knowing who the internal customers and suppliers are, and how they fit into the picture, is important, because together you form the 'end-to-end' process that ultimately provides the external customer with the service or product they're looking for. The external customer is someone outside of your organisation. They look to you to meet their requirements and pay you accordingly.

Consider Figure 3-2. Department A produces output for Department B, which produces output for Department C, which provides the answer to an external customer enquiry. Each of these departments is involved in the process and needs to understand the objectives of the 'big process'.

All too often, departments work in a vacuum, doing their own thing without regard for the impact of the end-to-end process. They may have their own targets, measures, and priorities, for example. Possibly, the end-to-end process isn't even known; each team or department involved works as though their step in the process is independent of any others. In reality the end-to-end process is a series of interdependent steps.

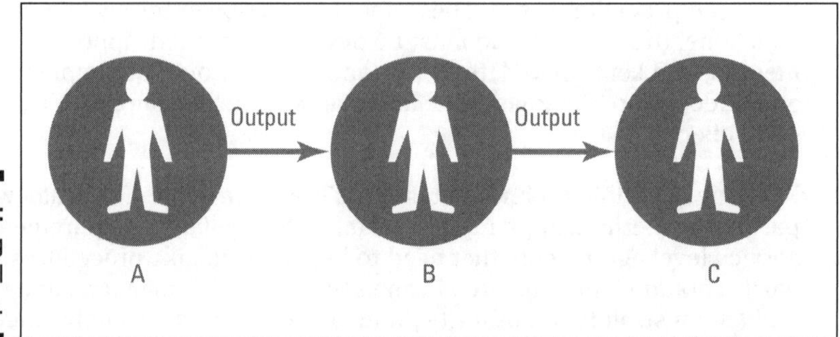

Figure 3-2: Identifying your internal customers.

Internal customers and suppliers must understand their relationship and their different roles. If they don't, the external customers will experience poor service – the very people who should be viewed as the most important because they're paying an organisation for the services or products they provide.

Even if you're not directly dealing with the external customer, you're quite likely to be dealing with someone who is. So, understanding the bigger picture is important, and meeting the requirements of your internal customers could well be the key to successfully meeting the external customers' CTQs.

Getting a High-Level Picture

To really understand how the work gets done, and to identify just who the internal and external customers are, you need to draw a picture of the process. These pictures are known as *process* or *value stream maps* (covered in detail in Chapter 5). Often, however, organisations map their processes in minute detail. You can get lost in too much information. Keep it simple.

Also, the pictures or process maps tend to be based on what people think is happening in their organisation rather than on the reality of the *Gemba* – 'the place where the work gets done' (see Chapter 5 for more on this concept).

Before developing a process map, recognising that different process levels exist in an organisation is important. Right at the top of an organisation are some very high-level processes, for example 'business development'. These Level 1 processes break down into a number of sub-processes. Level 2 and 3 processes gradually increase the amount of detail. You move from the high-level 'what' to the increasingly detailed 'how'. Levels 4 or 5 cover the step-by-step procedural activity.

Our example in Figure 3-3 has 'business development' at Level 1 and shows the various sub-processes down to Level 3.

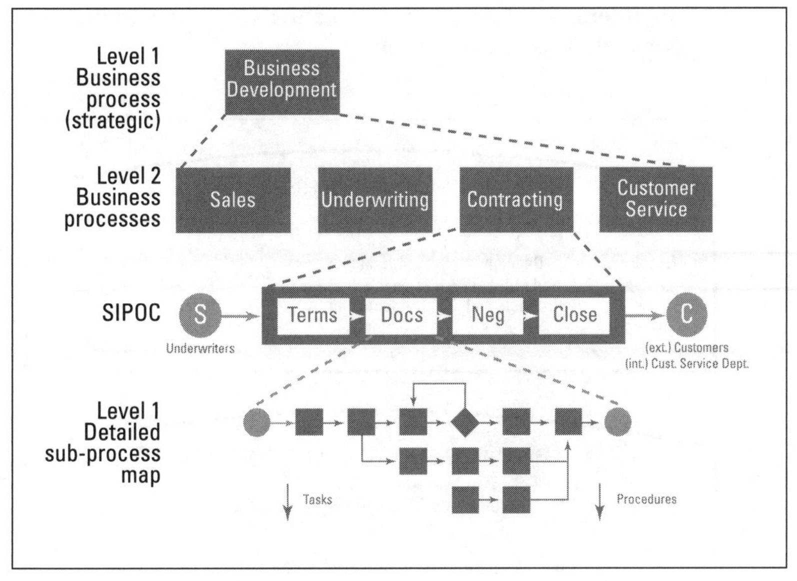

Figure 3-3: Process levels.

Figure 3-4 (overleaf) provides another example of Level 1 processes.

Drawing a high-level process map

A high-level process map provides a framework to help you understand your process and its customers and suppliers better, and to think about what needs to be measured in the process to help you understand performance and opportunities for improvement (covered in Chapters 6 and 7). Figure 3-5 shows the *SIPOC model*. SIPOC stands for:

- **Suppliers:** the people, departments, or organisations that provide you with the 'inputs' needed to operate the process. When they send you an enquiry or order form, the external customer is also included as a supplier in your process. Suppliers also include regulatory bodies providing information, and companies providing you with equipment or raw materials.
- **Inputs:** forms or information, equipment or raw materials, or even the people you need to carry out the work. For people, the supplier may be the human resources department or an employment agency.
- **Process:** in the SIPOC diagram, the P presents a picture of the process steps at a relatively high level, usually Levels 2 or 3.
- **Outputs:** a list of the things that your process provides to the internal and external customers in seeking to meet their CTQs. Your outputs will become inputs to their processes.

- **Customers:** the different internal and external customers who'll receive your various process outputs.

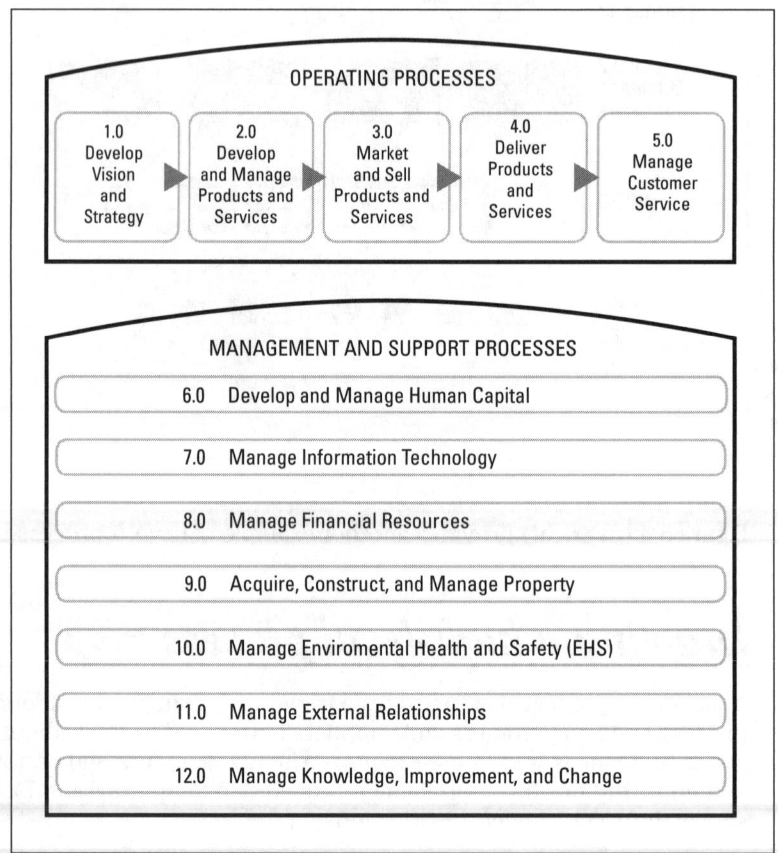

Figure 3-4: The APQC generic process framework

The SIPOC model identifies your customers and the outputs they need, presents a high-level process map, usually comprising four to eight steps, identifies your suppliers, and confirms your input requirements from them.

The customer can also be a supplier, particularly of information needed by you.

In many ways, SIPOC should really be called COPIS, because when you create the diagram you start with the customer on the right-hand side of the model, before listing the outputs that go to them. Well, you almost start with the customer. First, you and the process team need to agree a start and stop point for the process, so that everyone in the team is clear about the parameters.

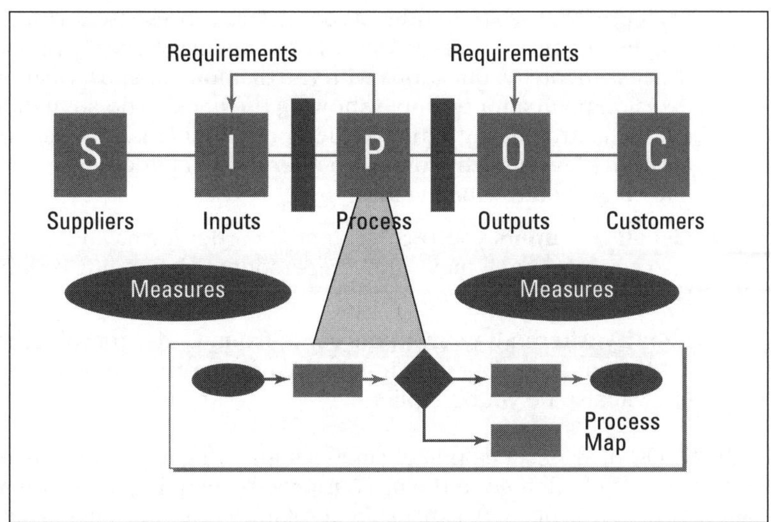

Figure 3-5: The SIPOC model.

For example, your process may start with the external customer sending you an order form for one of your products and end when the product has been sent to them or you've issued an invoice. Once the process parameters are agreed, you next consider the customer – which is easier if you have a clear understanding of their CTQs.

The best way to create your SIPOC diagram is to gather your team around a large sheet of paper and follow these steps:

1. **List all the different customers involved.** Include both internal and external customers, for example 'Management', who may need reports or information. Consider third parties, such as regulatory bodies, where relevant. Segment external customers by the different outputs they receive.
2. **List all the outputs you send your customers.** Now you have your list of customers, what is it that you need to provide them with? Under the 'O' for outputs, make a list of the things you send them. Drawing arrows showing who receives what output can be useful.
3. **Set out the steps in the process.** Use sticky notes to construct a high-level picture of the process. Typically, it involves four to eight steps. Don't go beyond eight steps as you'll be dealing with too much detail, too soon (Chapter 5 covers process and value stream mapping in detail, but the map starts here in the P of SIPOC.)

Involving the step-by-step procedures of a process in your map can make you lose your way in the various twists and turns that can occur. As a result, you may not be able to see the non-value-added steps. Spotting unnecessary activities is easier from higher up the map.

In Figure 3-5, the start and stop points are represented with the oval shapes, the process steps with the rectangular boxes, and points in the process involving questions with the diamond shapes. Diamonds are decision points, for example showing the need to do something different if you are dealing with product A or product B, or where different authority levels might come into operation in underwriting a loan, perhaps based on the loan value.

4. **List all the inputs you receive.** Include order forms and information, but also consider equipment or even personnel, depending on whether this is a new or existing process.
5. **Identify where all your inputs come from.** Under the 'S' for supplier, list the sources of all your inputs. Remember that some of your customers will also be your suppliers.

The SIPOC provides a helpful checklist, identifying who your customers are and the outputs that go to them. Your process map highlights areas where greater clarity is needed, especially in relation to requirements and outputs. It also helps you focus on what needs to be measured, for example, how well are you delivering the outputs to your customers, and how well are your suppliers meeting your requirements of them? (Measurement is covered in more detail in Chapters 6 and 8, with Chapter 7 focusing on how you present your data and understand variation in your results.)

Creating a SIPOC process map provides an opportunity to begin thinking about the various elements involved in your process, whether you have all the information you need, and if segmenting your customers is necessary.

Segmenting customers

In developing your SIPOC process map, you need to identify your customers and the outputs that go to them. Possibly, you classify or segment your customers in some way, for example by size or geographical location.

Think carefully about these different customers. Do they actually have different CTQs? Will the process outputs be the same for each segment, or will these vary to some degree?

We look at segmentation in a little more detail in Chapter 4, but ensure your SIPOC map and the thinking that accompanies it takes appropriate account of your different customer segments.

Chapter 4

Understanding Your Customers' Needs

In This Chapter

- Hearing the voice of the customer
- Putting your customers first
- Gauging how well you do

Throughout this book we make several references to the 'voice of the customer' and to CTQs (Critical To Quality requirements). The voice of the customer (VOC) helps you understand customer requirements. This expression describes information coming from the customer, perhaps through market research or face-to-face discussion, which enables you to determine the customer CTQs. The CTQs are vital elements in Lean Six Sigma, providing you with the basis to assess how well you're performing in meeting your customers' requirements. This chapter looks at how to obtain the VOC and develop the CTQs.

Obtaining the Voice of the Customer

We find out what our customers want by talking to, listening to, and observing them. Lots of sources of the customer's voice exist, such as market research results, focus group discussions, survey results, and customer complaints. The trick is to translate what the customer says into a measurable requirement – the *critical to quality* (*CTQ*) customer requirement. You gather input from your customers in order to understand their needs, identify the key issues, and translate them into terms that mean something to your organisation and that you can measure.

Listening to the VOC is about determining your customer requirements, not determining solutions to meet those requirements – it isn't about jumping to conclusions about what they mean, either!

Doing things properly usually involves a number of conversations with your customer, the process owner, and all the other people involved in the process. The *process owner* is the person responsible for the process, for example a manager or director; they need to ensure the process is designed and managed to meet the CTQs so they must understand what the customer is saying.

On occasion, the customer may not be totally clear about their requirements. Reflect back your interpretations to ensure they're correct.

Taking an outside-in view

Many organisations assume they know what their customers want – but this isn't always the case. Even when you try to be objective, the fact that you know your products and services so well, and understand your workplace's jargon, means seeing things from your customer's perspective is actually quite difficult.

Customers aren't all the same. They come in all sorts of different shapes and sizes, and your customers have different requirements, even for the 'same' product or service. Identifying the different customer segments that your organisation deals with, and recognising that each segment may have different CTQs, is essential. For example, a small company may be happy to receive your products on a monthly basis, whereas a larger business may need daily deliveries of those same products.

Segmenting your customers

Grouping your customers into segments helps you see your customers' different requirements. By segmenting them, you can develop the right products and services for each group, and create specific measures that help you understand your performance in meeting their differing requirements. To help you segment your customers, list some categories that describe both your current customers and the people or organisations that you consider to be potential customers.

Consider the following segmentation categories in relation to your customers:

- Industry
- Size
- Spend

- Geographical location
- End use
- Product characteristics
- Buying characteristics
- Price/cost sensitivity
- Age
- Gender
- Socio-economic factors
- Frequency of purchase/use
- Impact/opinion leader
- Loyalty
- Channel
- Technology

Naturally, you need to determine the categories that are relevant to your own organisation, but this list provides a good starting point. If appropriate, you can also create umbrella customer segments to include, for example, frequency of purchase and spend.

Prioritising your customers

Every customer is important, but some customers use your services more frequently or are more critical to your business than others. You may need to devote more time and resources to these particular customers.

Italian economist Vilfredo Pareto developed the idea of the 80:20 rule, when he described how 80 per cent of the wealth of his country was in the hands of 20 per cent of the population. Your organisation may well have a 'vital few' customers, perhaps 20 per cent who provide you with 80 per cent of your business. If that's the case, those customers are very important to the ongoing success of your business and understanding their requirements and their perceptions of your performance in some detail is crucial.

You could prioritise by customer segment, but you need to understand how best to prioritise your customers for your organisation. For many organisations, the priority may be revenue, but more sensibly it should be on profit. As your understanding of your customers increases, you'll find it easier to determine your organisation's priorities.

Truly focusing on customers, as opposed to simply saying you are, requires real investment in understanding your customers' needs. Knowing who your customers are, how they're segmented, and which ones are your priority is a vital prerequisite to your research and data gathering. Identifying the wrong customers, or not being aware of the different segments, can mean you collect information on requirements or customers that aren't related to the process or service you're designing.

Your customers may be both internal and external. Thinking in terms of processes helps you identify where you need to focus, and highlights who your internal customers are.

Researching the Requirements

Researching your customers follows a natural progression. You start with potentially no information about customers, and end up with a collection of quantified, prioritised customer needs and expectations. You might also gain information about how your competitors succeed in meeting their – and your! – customers' needs.

Start by investigating what information you already have and then determine the 'gaps' in your customer information. For example, if your proposed product or service is a variation on something you've done before, you may have a lot of information in your organisation's files. Likewise, if one of your competitors already offers the product or service, you may find some useful industry information.

You then need to develop a customer research plan that moves you from where you are right now to where you need or want to be – you close the 'gaps'. Use Table 4-1 to help you determine the sequencing of your research plan.

Table 4-1 Researching the Requirements

Input	*Research Method*	*Output: What You Get*
No information	Interview/focus group What is important?	Customer wants and needs (general ideas, unprioritised, not clarified, all qualitative)
Known preliminary customer wants and needs	Interview/focus group Which are most important?	Customer wants and needs (clarified, more specific, preliminary prioritisation) Customer input to list of competitors, best-in-class

Input	*Research Method*	*Output: What You Get*
Qualitative, prioritised customer wants and needs	Survey Face-to-face Written mail Telephone Electronic	Quantified prioritised customer wants and needs Competitor comparative information

Table 4-1 helps you think about the information you currently have, and the information you need to get and how to go about getting it.

If you have no information to begin with, the first step is interviewing some representative customers, perhaps in a focus group, to help you understand what's important to them in the service or product you're offering. You should be able to capture some general ideas about their wants and needs; though these are unlikely to be totally clear, and you probably won't be able to prioritise them yet. This first step helps you determine your customers' preliminary needs and enables you to now run some more detailed interviews or focus groups.

The second row in Table 4-1 shows how more detailed analysis should lead you to a clearer picture of customer requirements, and at least a preliminary prioritisation of them. This stage also gives you the opportunity to ask the customers about their experience of your competitors.

With this clear picture, you can move to the third row in the table. Your interviews and focus groups will have been run with a relatively small number of customers. This approach is described as *qualitative* research. You now need to test their views and opinions by conducting research with a larger number of customers. This approach is described as *quantitative* research, involving a survey.

Quantitative research is important – it enables you to feel confident that you have a true picture of customer wants and needs. The results from the qualitative research may have been skewed in some way if you'd inadvertently included unrepresentative customers. For example, you may have carried out research on a product aimed at the high end of the market, but interviewed customers from a different market segment by mistake.

As you plan your research, be aware of the following issues:

- The customer may offer you a 'solution' rather than express their real needs. Ask the customer 'Why do you want this?' until you truly understand the real need.

By asking its customers to express their real wants, Xerox refocused its entire business on its customers' need for documents rather than just photocopiers.

- Different customers may perceive the same product or service differently. For example, a shirt that shows off the designer's logo may command a premium price over a similar article without the logo – but only for some customers.

Ford offers the 'Mercury' version of many of its cars to appeal to its more luxury-conscious customers, while Toyota uses the slogan 'In pursuit of perfection' to describe its Lexus brand.

- Remember that how your customers say they'll use a product or service isn't always the same as how they actually use it!
- External customers generally express *effectiveness needs* – needs that relate to the value the customer receives from the product or service. Internal customers, on the other hand, tend to express *efficiency needs* – needs that relate to the amount of resources allocated or consumed in meeting customers' needs.

Interviewing your customers

Customer interviews are helpful as a research technique. The aims of customer interviews are to *understand* a specific customer's needs and requirements, values and points of view on service issues, product/service attributes, and performance indicators/measures. Customer interviews are useful:

- To explore issues with customers during customer research.
- At the beginning, to learn what is important to customers which supports the development of hypotheses about customer values.
- In the middle, to clarify points or to better understand why a particular issue is important to customers.
- At the end, to clarify findings, to get ideas and suggestions, or to test ideas with customers.

The advantages of customer interviews are:

- Flexibility – you can obtain more detailed explanations by probing and clarifying.
- Greater complexity – you can administer highly-complex questionnaires/ surveys and can explain questions to interviewees.
- Able to reach all customer types – you can interview populations that are difficult or impossible to reach by other methods.

- High response rate – the degree to which the information collection process reaches all targets is higher.
- Assurance that instructions are followed – because the interview is taking place in person, you can ensure that all steps are followed.

The disadvantages of customer interviews are:

- They're costly to administer.
- They're the least reliable form of data collection because the interviewer may influence the responses to the questionnaire.
- Less anonymity is possible.
- Interview time is limited to 15–20 minutes (business-to-business customer interviews are generally 45–50 minutes' duration).
- Generating supportable quantitative evidence can be difficult.
- Results can be difficult to analyse.
- The sample size may not be sufficient to draw supportable conclusions.
- Use of different interviewers asking questions in a particular way may result in bias (see the 'Avoiding Bias' section later in this chapter).
- Positive response bias may occur, whereby people give higher ratings in personal interviews.

Ask open questions to get the interviewee talking, rather than asking a series of closed questions that simply elicit a yes or no response. The key is to listen to what the customer is saying – their responses often provide the answers to some of the specific issues you need information on.

Focusing on focus groups

Focus groups are interviews, usually involving between six to ten people at the same time in the same group. Typically, they run for two to three or four hours. Focus groups are a powerful means to capture views and opinions, to evaluate services or test new ideas, or (in the context of this chapter) to clarify and prioritise customer requirements.

Essentially, a focus group is a carefully planned discussion designed to obtain perceptions on a defined area of interest in a non-threatening environment. Listening to the members of the focus group is key!

The focus group participants should share characteristics that relate to the focus group topic, all being in the same customer segment, for example. To avoid bias, running more than one focus group is sensible; run three as a minimum.

Focus groups are used:

- To clarify and define customer needs.
- To gain insights into the prioritisation of needs.
- To test concepts and receive feedback.
- As a next step after customer interviews or a preliminary step in a survey process.

The participants will be asked to thoroughly discuss very few topics, and often only three or four questions will be asked during the focus group. These will be very general questions, such as 'what do find important in service delivery, generally?' or 'what is it that makes you feel you have received good service or bad service?'

Focus groups aim to get the participants talking and you listening, ideally recording the discussion on video or tape for subsequent analysis. Table 4-2 lists the advantages and disadvantages of using focus groups.

Table 4-2 Pros and Cons of Using Focus Groups

Use Focus Groups When:	*Don't Use Focus Groups When:*
You need to make or confirm market segmentation decisions	The environment is emotionally charged and more information of any type is likely to intensify the conflict
Hypotheses about the market and customer values need to be developed or tested in exploratory or preliminary studies	Highly valid, quantitative data is needed
A communication gap appears to exist between your company and the market segment	Other methodologies can produce either better quality or more economical information
Insight is needed into customer perceptions over potentially complicated topics where their opinions and attitudes may influence your course of action, perhaps in launching a new product	The researcher cannot ensure the confidentiality of sensitive information

Use Focus Groups When:	***Don't Use Focus Groups When:***
Synergy among individuals would be useful in creating ideas	You're trying to sell products
Hypotheses need to be developed in preparation for a broad survey or large-scale study	
A higher value is placed on capturing open-ended comments than data from the target audience	

Considering customer surveys

You can run customer surveys in a number of ways, including by postal questionnaire, or electronically by email or the Internet. They enable you to measure the importance or performance of customer needs and requirements. You can check out your focus group findings with a larger group of customers.

The pros and cons of customer surveys are shown in Table 4-3.

Table 4-3 Pros and Cons of Customer Surveys

Advantages	***Disadvantages***
Low cost	Low rate of return
Efficiency of large samples	Non-responsive bias
Ready access to hard-to-reach respondents	Little control
No interviewer bias (though beware of the questions)	Limitations on questions
Potential for exhibits	Potential misunderstanding of questions or rating scale
High reliability and validity	Oversimplification of format
	Slowness – requires development
	Requires pre-testing
	Difficulty of obtaining names

Using observations

Observations are another way of identifying your customers' needs and CTQs. Observing a customer is an effective way to understand how that customer uses and views your products and services. For example, by observing the purchasing patterns of customers, supermarkets can strategically position the products in their store to increase sales. Key products appealing to elderly customers are positioned on middle shelves so they don't have to bend or stretch to reach them. The CTQs for this particular customer segment have then been met.

Toyota allocates some of its engineers to ride as passengers in customers' cars, enabling them to observe their customers' driving first hand. One result of such observations was the introduction of drink holders in Toyota's cars. The engineers observed children in the back seats holding drinks but having nowhere to put them. The inclusion of drink holders may not have increased sales as such, but they're now a standard requirement for most buyers, and their inclusion in a car as 'standard' helps maintain or enhance the general level of customer satisfaction.

Avoiding Bias

Whichever approach you take to collecting voice of customer (VOC) information (see the earlier sections in this chapter), you need to recognise the potential for bias. Possibly you may ask the wrong questions, or ask the wrong customers because you haven't segmented them properly, or simply misinterpret, deliberately or otherwise, what the customer has said.

You really need to listen to what the customer says – not what you think they're saying or you'd like them to say! In a focus group, asking very open questions is best and then listening to their responses. Leading customers through a series of closed questions results in closed answers, which are then open to interpretation and bias.

On questionnaires, the wording of questions is vital. Each question should ask only one question. So, for example, you might ask the customer to rate your performance on a scale of one to five, where one is poor and five is super. If you pose a question such as, 'How well do you feel we perform in terms of speed and accuracy?', you may receive a response relating to only speed or accuracy, not both.

In the fictional British television series *Yes, Prime Minister* there's a great example of how asking closed questions can lead to two different answers to the question: 'Would you be in favour of reintroducing National Service?' In the first survey, the questions include 'Are you worried about the number of young people without jobs?', 'Are you worried about the rise in crime among teenagers?', and 'Do you think young people welcome some authority and leadership in their lives?'

Given the previous questions, the respondent is almost obliged to say yes to reintroducing National Service. The bias is increased if the results are published without reference to the previous questions.

Asking a different set of questions elicits the opposite response. Questions include 'Are you worried about the danger of war and the growth of armaments?', 'Do you think there's a danger in giving young people guns and teaching them to kill?' and 'Do you think it's wrong to force people to take up arms against their will?' This time the respondent can only answer that they *oppose* the reintroduction of National Service.

Beware of the huge potential for bias, be it innocently introduced or deliberately created.

Considering Critical to Quality Customer Requirements

When you have collected the VOC information (see the earlier sections of this chapter), you need to develop the CTQs. Write the CTQs in a measurable form: they provide the basis for your process measurement set. This set will enable you to put the right measures in place to assess your performance (see Chapter 6).

CTQs help you focus on your customer requirements and provide the foundation for your measurement data. Figure 4-1 provides a framework to help you define your CTQs. You can identify two key issues – getting through to the right person and speed. Looking at the first example, the CTQ for getting through to the right person first time is straightforward to understand and to measure, but to define the CTQ covering speed you need to go back to the customer and agree what 'quickly' means. You might then define a second measurable requirement as 'The call is answered within 10 seconds.'

Without this type of data, you won't know how you're performing in meeting the customer requirements – information that determines where improvement actions are required.

A CTQ shouldn't prescribe a solution. A CTQ should be measurable and, where appropriate, have upper and lower specification limits and a target value. A CTQ should be a positive statement about what the customer wants rather than a negative statement about what the customer doesn't want.

Say you agree a time window for a customer's boiler to be repaired. The upper specification limit might be midday, the lower specification 8 a.m, and the target time 10 a.m., for example. So you aim to be there for 10 a.m. plus or minus two hours.

Voice of the Customer	Key issues(s)	CTQ
You either put me on hold, or put me through to the wrong department or person	The customer wants to be put through quickly to the right person	• Customer gets through to the correct person the first time
You send me an invoice at different times of the month	Consistent monthly billing	• Customer bill received same day of the month
It takes too long to process my mortgage application and get me the money when it's needed	Speed up loan so I get the money on time	• Customer receives cheque on customer request date

Figure 4-1: Determining the CTQs.

The affinity diagram (see Figure 2-3 in Chapter 2) provides a useful format for sorting VOC information into themes. These themes can then be broken down into more detailed elements, as a CTQ tree, as shown in Figure 4-2.

First level	Second level	
Friendly staff	Willing to answer question	*You will probably need to go down to more levels. And remember the CTQs need to be measureable*
	Shows respect	
Knowledgeable staff	Knows the loan process	
	Knows the market	
	Understands my situations	
Speed	Money when I need it	
	Application fast to fill out	
Accurate	Don't make mistakes	
	Give me the right rate	

Figure 4-2: Developing a CTQ tree.

The example in Figure 4-2 is from a bank that has taken the various customer statements and comments from a survey and sorted them into themes using an affinity diagram. The high-level themes are shown in the first level; the second level shows some of the comments within the theme. These comments break down into another level of detail (rather like the branches on a tree), to ensure the requirements are properly understood – we thus have a 'CTQ tree'.

When you develop CTQs, you can usually group customer requirements under common sets of headings. We do this, and show a selection of examples and potential measures, in Table 4-4.

This common list allows you to structure the process of gathering requirements and reduces the risk of you missing a CTQ.

Table 4-4 Some Common CTQs

CTQ grouping	*Examples*	*Measures*
Speed	Bills paid on time (in and out)	Elapsed times
	Deliveries made on time	Turn-around times
	Time to answer calls	Call answer rate
	Turn-around time on IT project delivery	Call abandon rate
Accuracy	Orders containing the correct information Computer system that works	Number of defects in orders, deliveries, products, or software Number of calls to helpdesk Number of bugs reported by users testing a new computer system or a program change
Capacity	Needs to cope with the right volume of orders/number of simultaneous enquiries	Number of items per order Number of clients Number of concurrent users Number of orders per day

(continued)

Table 4-4 *(continued)*

CTQ grouping	*Examples*	*Measures*
Data/ information	Easy access to order details and status Software developer needs to understand how a software module works before changing it	Salesperson can access order details and status within five minutes of request while on the road Time spent converting data and cleansing data Percentage of software modules not meeting development standards
Safety	No customers or staff injured on company premises	Score on monthly safety audit
Compliance	Data Protection Act Consumer Credit Act Health and Safety Act	Complies
Security	Very difficult to defraud the company either from within or externally	Customers' or suppliers' risk rating Score on six-monthly security audit Maximum Risk Priority Number (RPN) rating from a Failure Modes Effects Analysis (FMEA) – a prevention technique covered in Chapter 10 Time since last FMEA performed
Money	Cost of supplies Commissions paid Limits on size of order	Ratio of size of order in cost terms to customer's risk rating

CTQ grouping	*Examples*	*Measures*
People	Staff needs to be adequately trained Need access to key staff to assist with requirements, design, testing, and so on All functions properly consulted on requirements and participate fully in requirements review and design reviews	Approval review scores on requirements documents Approval review scores on design documents Scores employees achieved on post-training exams Hours worked on project with day job Employees' scores on quarterly HR questionnaire
Professionalism	Integrity Knowledge Courtesy Availability	Scores on knowledge tests Score on post-delivery follow-up call to customer Scores on HR psychometric tests Number of calls not answered first time by the first point of contact Call answer rate Call abandon rate
Social conscience	Ethnic mix Equal salary structures for men and women Contributions to charity	Percentage of turnover given to charitable causes
Environment	Recycling policy Biodegradable packaging Emissions	Percentage of offices with recycling containers
Changes/re-schedules to original order or contract	Conflicting requirements between customer and supplier	Ratio of cost of change to extra charge to the customer

Establishing the Real CTQs

Interpreting a customer's stated CTQ as a real CTQ that can be measured presents a big challenge. Often customers jump to preconceived solutions and prescribe those solutions as part of their requirements. If you take these customers' requirements literally, several problems can occur, including missing the customer's real requirement. The product or service you provide them with may then not be quite right. In turn, misunderstanding could then lead to you delivering a more expensive or less efficient solution than the particular CTQ requires.

The secret to finding the real CTQs is to keep challenging the customer by asking 'Why?' until the need falls into one of the categories in Table 4-4 or is otherwise clear. Below are two, disguised, real-life examples:

- An internal customer said, 'We need one integrated SAP system handling all orders instead of splitting orders between our different European divisions.' But *why* do we need this?

  The internal customer responded: 'Because customers think we are unprofessional.' But *why* do customers think that? The internal customer's answer explained that customers get more than one order acknowledgement if the order is split between divisions.

  This answer gives us the real CTQ: 'Customers require single order acknowledgement for all orders.' In Table 4-4, this CTQ comes under the Data/information category. You may find several solutions to meet this requirement without going to the expense of a single integrated SAP system.

- An internal customer asks for a web-based order enquiry system. But *why* do they need this? They respond that enquiries currently take ages.

  But why do enquiries take so long? The current process involves having to go to four different screens to get the information needed. By asking why speed is important, we discover that the customer is left waiting on the phone, but they expect the answer within 30 seconds.

  The real CTQ becomes: 'Customers' order enquiries by telephone should be satisfied within 30 seconds', which comes under the Speed category in Table 4-4. As with the first example, there could be several solutions to meeting the CTQ – the web-based idea may not be the most appropriate or the most economical.

Prioritising the requirements

Clarifying your customers' CTQs (which we describe in the previous section) is vital, but you also need to find out which of the CTQs are especially important.

You can prioritise your CTQs in a number of ways. You can simply ask your customers to weight their own CTQs, or you can use a simple tool such as *paired comparisons*.

The paired comparisons technique provides a way to determine priorities and weight the importance of criteria. Using the paired comparisons tool forces you to make choices by looking at each pair from a list of options – in this case, a list of CTQs. Instead of asking your customers to identify their top choice, you ask them to select their preference from each pair.

For example, if you have five CTQs, you ask: 'Do you prefer A or B? A or C? A or D? A or E?' After A, you compare B and C, B and D, and B and E, and so on.

You can use this technique face-to-face, over the phone, or by using a 'voting grid' like that shown in Figure 4-3, where participants circle their choices for each comparison.

Item	Description				
A		A/B	A/C	A/D	A/E
B			B/C	B/D	B/E
C				C/D	C/E
D					D/E
E					

Figure 4-3: Paired comparisons: Do you prefer this or that?

Measuring performance using customer-focused measures

Talking about being customer-focused is much easier than actually *being* customer-focused. Using *outside-in* thinking and measures to assess your CTQ performance is one way to help you think differently and focus on the customer.

Determining your CTQs provides a basis for your measures. In Chapter 6 we look at measurement and data collection in some detail, but here we take a brief look at some of the different thinking you need in order to focus on your customers.

Think about what your customer sees and experiences in terms of your organisation's products, services, and performance in meeting customers' requirements. Consider, for example, whether being a customer of your organisation is easy. Many organisations are internally focused and think negatively of their customers – but ultimately customers pay the bills.

Try to drag yourself outside your organisation and take a look in – 'outside-in' thinking. Think about what your customers see and consider whether they're happy.

Understanding what your customers measure is helpful. Ask yourself whether your customer measures the same things as you – and then think about how their data compare with yours. Consider why differences may be evident. Then think about what your customers do with the output from your processes: where it fits in their processes. Here's a real-life example.

Airlines make money when their planes are in the air. When a plane is out of commission, perhaps for servicing, the airline makes less money – the company needs the plane up and flying again as quickly as possible.

General Electric's (GE) aircraft engines division discovered the value of outside-in thinking when they realised their customers were measuring GE's performance a little differently from the way they did it. GE would receive an engine for servicing into their process; and their clock would start. When the service was complete, their clock stopped and they reported that this service had taken x hours to complete.

What they were forgetting was the fact that their customers were counting the time from when the engine came off the plane to the time when it was put back on – the *wing-to-wing time*. The phrase and the thinking caught on. Chief Executive Officer, Jack Welch, deployed this concept throughout the GE operations and divisions, worldwide.

Think about how your measures measure up. Is there scope for wing-to-wing thinking in your processes?

Chapter 5

Determining the Chain of Events

In This Chapter

- Following a chain of events from start to finish using process stapling
- Drawing a spaghetti diagram to see how the work gets done
- Creating a map of the process

As a manager, your role is to work on the processes that you manage with improvement in mind. You therefore need to know precisely how these processes work. Having an up-to-date picture of how things are done makes DMAIC (Define, Measure, Analyse, Improve, and Control) improvement projects far easier to undertake (dive into Chapter 2 for more on doing the DMAIC).

The 'Measure' phase of DMAIC is about understanding how and how well the work gets done. We look at the 'how well' aspect in Part III; our focus here is to understand how the work currently gets done. Only after you understand how the process works *now* can you see the opportunities for improvement in your process and manage performance better.

Finding Out How the Work Gets Done

How to draw a process map is the main focus of this chapter. We look at two types: the deployment flowchart, and the value stream map. These maps build on the high-level SIPOC diagram explained in Chapter 3, and provide really helpful pictures of how the work gets done.

Before you draw any kind of process map, visit the workplace and see for yourself what's really happening. The Japanese refer to this observation as 'going to the Gemba'.

Genning up on the Gemba

'The Gemba' is a Japanese term for the 'actual place' – that is, where the action is. Only in the Gemba can you truly see how things are done and it's the only place where real improvement can occur. You may be able to draw up new ways of doing the work in some central management location, or in an engineering office, but the reality is the Gemba. That's where things are defined, and refined, to produce genuine and effective change.

You're likely to find surprises waiting for you in the Gemba. Very often you'll find the process is being carried out differently to how you thought it was happening, especially when more than one team is involved. We cover techniques such as *process stapling* and *spaghetti diagrams* in this chapter. These techniques help you to see the reality of your workplace and enable you to identify unnecessary steps and eliminate waste. (We wade through waste in detail in Chapter 9.)

Practising process stapling

Process stapling offers one way to really understand the process and the chain of events. Very simply, *process stapling* means taking a customer order, for example, and literally walking it through the entire process, step-by-step, as though you were the order.

No matter where the order goes, you go too. By following the order you start to see what really happens, who does what and why, how, where, and when they do it.

Carrying out a process stapling exercise with a small team of people can be an ideal first step. Sometimes, there can be advantages in beginning the exercise from the end of the process and working backwards. People will be less familiar with this 'reverse flow', helping them think more carefully about things.

The nearby sidebar 'Seeing process stapling in action' demonstrates the power of this technique. You begin to understand all the steps in the process and how much time and movement is involved in carrying out the work. Process stapling helps you identify a number of improvement opportunities, even if you don't use the exercise to create a spaghetti diagram or process map.

You might, for example, spot the scope for tidying up the workplace, making it easier and safer to find things. (We shine a light on neatness in Chapter 10.) The process stapling exercise helps you spot the frustrations in the process, such as inconsistencies and why-on-earth-do-we-do-this? activities. You can then spot the steps that add value and those that don't. (We unveil value in Chapter 9.)

Process stapling in action

This example reflects our experience of process stapling in at least one of our client organisations. Ann receives a customer order – she needs to input some information to the system, print out an internal form, add some additional information to it, and then send it to Brian.

You now need to 'staple' this form to yourself and take it over to Brian (imagine attaching it to your clothing, for example). Brian is some distance away. Immediately, you have a sense of how much transport is involved.

When you get to Brian, you find that his first action is to correct all of Ann's mistakes. You ask Ann whether she's aware that she'd got things wrong. She's not happy about this, as she thinks she'd been doing what Brian needed, and had always done it this way. Ann tells you that Brian has never mentioned anything about her mistakes.

You find that Brian never bothers to tell Ann about the errors, which had been caused by misunderstanding, because he finds it easier to correct the mistakes himself.

After Brian corrects the errors, he sends the papers to Clare. You're dismayed to find that Clare sits next to Ann – shame the papers didn't go straight to Clare in the first place!

Clare tells you that this step is a complete waste of time. She's told her manager this, but her manager says the step's an important element of Clare's work. Clare just checks that the system is updated, that certain information is in the right box on the form, and that Brian has put his signature on the form. She finds this task boring and has never yet found a case that needs correction, so she simply puts these items to one side, lets the work build up, and then clears them all on a Friday afternoon before going home.

Typically, only 10–15 per cent of the steps in a process add value, and more often than not, the 'thing' going through the process spends as little as 1 per cent of the total process time in these steps.

When introducing the idea of process stapling, you may find some people telling you that this is what they already do. But what they actually do is get a group of people in a room and use sticky notes to help draw up the process. They're missing the point! The picture they draw will be what they think is happening. Process stapling enables you to see what's really happening.

Try taking photos of each step in the process. Apart from providing an ideal record of what you've seen, photos enable you to make an effective presentation to management of what you've found. Be prepared for them to be surprised. You can have fun taking the pictures (but not of the managers falling off their chairs in shock, however!), especially if you act the role of the thing going through the process.

As your understanding of the process increases, you're likely to find real value in working with your customers to extend the process stapling concept to incorporate their activities with yours. In this way, you can work out how your process and its output link to your customer's process, what your customer's process looks like, and how your customer uses your process outputs.

Extending process stapling provides great insight into how you can generate improvements in your process that really add value to your customer and make an impact that delights them. The technique can also lead to joint improvement activity with a DMAIC project being carried out in concert with your customer.

Drawing spaghetti diagrams

A spaghetti diagram provides a picture of what's happening in the process in terms of movement. The diagram tracks the movement of the thing or things going through the process, including the flow of information and the people carrying out the work.

In Figure 5-1 we show a pretty confused series of movements in a garage as an example. You can apply the technique to any working area, including your office or even your home. Think about the movements you make and the distance you travel when undertaking tasks such as making photocopies, picking up your printing, or making a cup of tea.

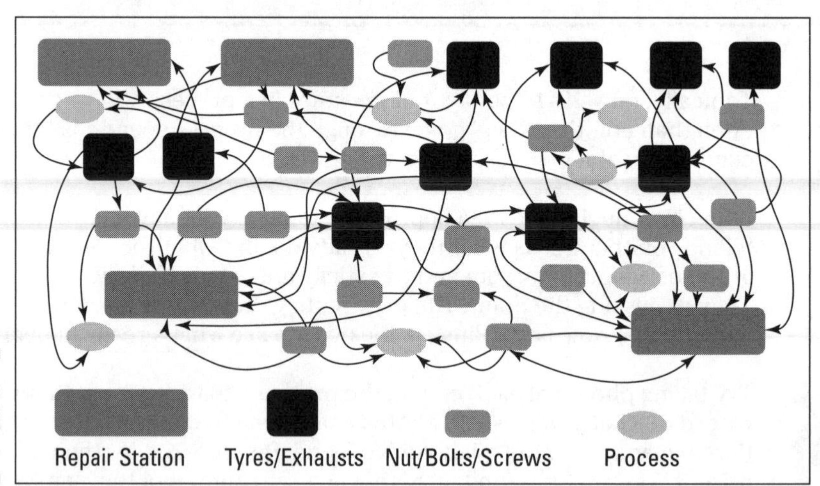

Figure 5-1: A spaghetti diagram.

The spaghetti diagram may throw up some real surprises about how much movement happens in your organisation, including how often things go back and forth. This technique helps you identify waste and provides a visual catalyst to stimulate change in your workplace.

You may already have used this technique at home! If you've installed a new kitchen, you'll know the importance of the triangle formed by the sink, cooker, and fridge.

The different shades you use in Figure 5-1 have no special significance, but you do need to distinguish between the movement of people, materials, and information. When you create a spaghetti diagram, you may want to use a current office plan, for example showing where the furniture, equipment, and power points are located. Make sure the plan really is current and that it includes all additional items, including those boxes in the corner that seem to have appeared from nowhere.

In developing a spaghetti diagram you can measure how far and why people are moving. You may be able to make some simple changes to your office layout to reduce the distance moved, or even to avoid it completely. You could even use a long ball of string or a pedometer to help you develop a more accurate diagram and better understand the movement involved.

When you use process stapling and spaghetti diagrams together, you may see the opportunity for a significant reduction in wasted movement and in other non-value-added steps too. So, in the process stapling example described in the nearby sidebar, is Brian's step necessary? If it is, would sitting Ann, Brian, and Clare more closely together make sense?

Unnecessary travelling and movement waste so much time. Siting the relevant people and equipment together is often a relatively simple way of reducing waste and processing time.

Painting a Picture of the Process

When trying to understand your processes and how the work gets done, the phrase 'a picture paints a thousand words' is certainly true. In this section we look at two specific options for painting that picture of your process – a deployment flowchart and a value stream map. Despite their names, these are both 'process maps', and we tend to use the term 'process map' in this section.

When you paint the picture of your process, keep in mind why you are doing it. Developing the picture helps you understand how the work gets done and the degree of complexity in the process. Your picture can highlight the internal and external customer and supplier relationships or 'interfaces', and help you determine the input and in-process measures you need (see Chapter 8 for more on these).

You're not painting this particular picture as the specification for a computer system change, so keep things simple. This picture is for you and will help you manage and improve the process. You are drawing a 'current state' picture to see how things are done *now*.

A 'future state' map shows how the process could be undertaken to achieve a higher level of performance at some future point. Achieving that performance may be harder and could result in the need for a DMAIC project (see Chapter 2 for a detailed overview of DMAIC). When you've drawn and implemented your picture of a future state map, it becomes the current state map. With continuous improvement in mind, you now need to develop a new future state picture.

Your picture can provide a useful framework that prompts a whole range of questions:

- Who are the customers that have expectations of the process?
- Why is the process done? What is its purpose? Does everyone involved understand the purpose?
- What are the value-added and non-value-added steps?
- How can you carry out essential non-value-added steps using minimal resources?
- What are the critical success factors – that is, the things you must do well?
- Why is the process done when it is done?
- Why are tasks in the process carried out in that order? Are all the steps involved in the process necessary? Do all the steps add value for the customer?
- Why is the process carried out by a particular person or people?
- What measurement is in place to assess performance and identify possible improvement opportunities? Think in particular of how you might identify and measure those parts of the process that are repetitive and important to ensuring the process conforms to requirements.
- What is the cycle time involved in the process? Why is the cycle time longer than the unit time?
- What are the barriers that prevent the supplier from producing a quality output?
- If decisions need to be made as part of the process, are the criteria that will be used to make the decisions understood by everyone involved? Are the decisions communicated adequately? Are the authority limits appropriate?

- How do you and others deal with problems that occur in the process?
- What are the most common mistakes that occur in the process? What impact do these mistakes have on the customers?
- Where have improvements already been tried in the process? What was the outcome?

Whichever questions you ask, don't forget to keep asking 'Why?'

Keeping things simple

Process mapping uses lots of different symbols, or 'conventions'; try to use as few as possible. To create a deployment flowchart, which we talk about in the next section, just two or three conventions are usually enough: the circle, the square box, and the diamond, as shown in Figure 5-2:

- The circle indicates the start and stop points in your process.
- The square box signifies a step or action.
- The diamond poses a question, where the answer determines which route the process follows next.

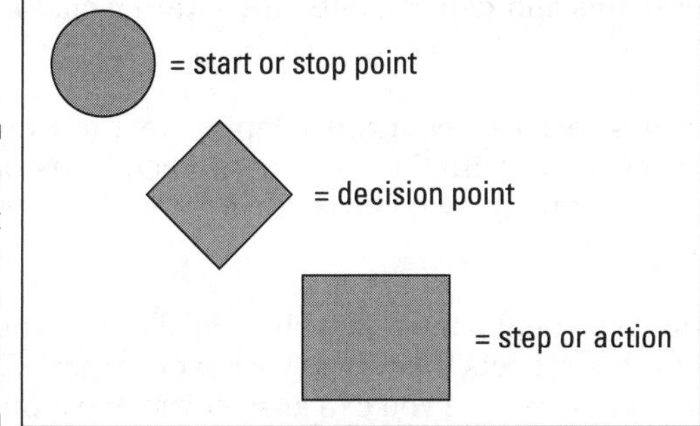

Figure 5-2: Keeping it simple with process mapping symbols.

Take a bank underwriting loan application as an example. The process steps may be different depending on the amount of money being requested as a loan. In the case of underwriting the request, it may be that large cases need to go to a senior underwriter, or require key documents from the client, whereas a small loan might be processed at a more junior level, or need less documentation. So, the diamond indicates a decision point with a question about the size of the loan, as shown in Figure 5-3.

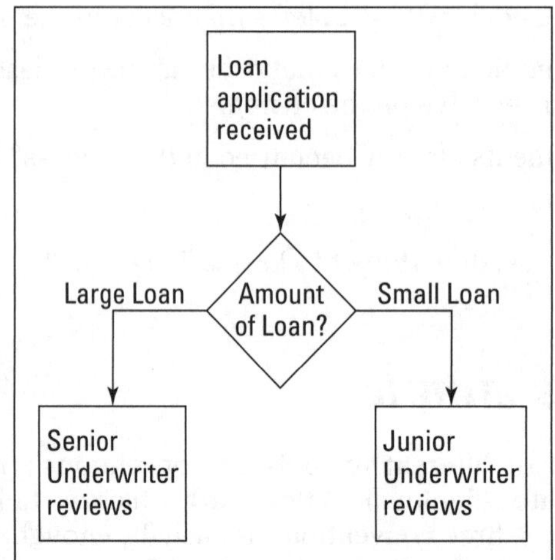

Figure 5-3: Which route do we follow?

Developing a deployment flowchart

The deployment flowchart builds on the high level SIPOC diagram described in Chapter 3, and goes into a little more detail, but not too much. This flowchart identifies who's involved in the process and what they do, including the different members of a team who are involved in different stages of the process, and also other teams and departments, the internal customers, and suppliers.

Spotting *moments of truth* is easier when using a deployment flowchart. Moments of truth are *touch points* with the customer (when a customer comes into contact with a company), which we consider in the later section 'Identifying moments of truth'.

Before you begin working on a deployment flowchart, make sure you have an objective for the process that reflects the CTQs (we cover Critical To Quality elements in Chapter 4). And make sure you can answer the question 'Why are you doing this process?'

Involve the people who work in the process when you develop a deployment flowchart. Because different perceptions exist on how the process works, use a sticky note for each step in the process so that you can move things around simply. You may well discover that the process is more complex than you think it is, which is why carrying out a process stapling exercise first can be so useful.

When you've used your sticky notes to create a flowchart, consider using a process mapping software to formally document the process. Many packages are available on the Internet, some offering free 30-day trials, so you can try things out and judge their suitability.

In the sidebar 'Seeing process stapling in action' earlier in this chapter, we introduce Ann, Brian, and Clare. If you haven't read this sidebar, have a quick flick through it – we use the same example, beginning with Figure 5-4, when we develop our map. Each time a different person or another area in the workplace is involved, the chart moves horizontally and down.

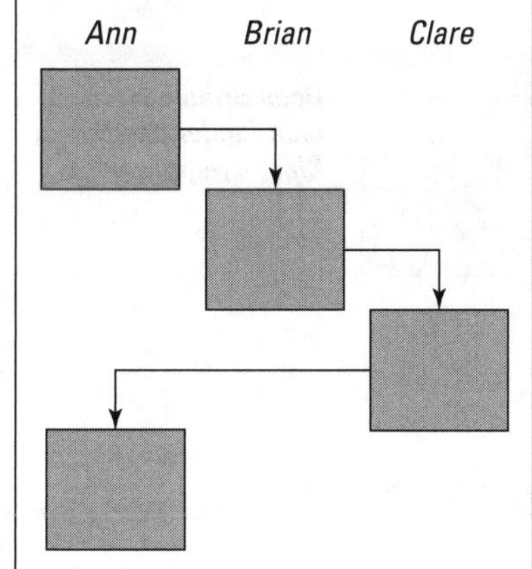

Figure 5-4: The deployment flowchart.

In Figure 5-4, we've simply focused on Ann, Brian, and Clare, but in practice you need to include the customer in the picture to help you identify the moments of truth (see the 'Identifying moments of truth' section, later in this chapter). Computer systems can also be included in your cast of characters. Work might be input to the computer system, for example, with the output coming out somewhere else. Seeing the whole picture is vital.

We cover measurement in more detail in Chapters 6, 7 and 8, but here we highlight some of the opportunities to put measurement in place. So, for example, when the chart moves horizontally and down, a customer and supplier relationship exists, as highlighted in Figure 5-5.

Most problems occur at the interfaces between two people or two departments, for example between Ann and Brian. Measures are almost certainly necessary here to help monitor performance and identify if problems exist, perhaps caused by misunderstanding the requirements.

Chapter 7 looks at the need to collect good data and develop a data collection plan, and Chapter 8 considers the importance of 'in-process' measures. Your results here, especially the level of rework, will have a major impact on your performance for the customer, so gathering good data and knowing what's happening is essential. In Figure 5-4, for example, you'd want to know whether Ann's output to Brian and Brian's output to Clare is always correct, and so on. If it isn't, you'd then want to find out what type of errors were occurring so that you could begin the process of improving the situation.

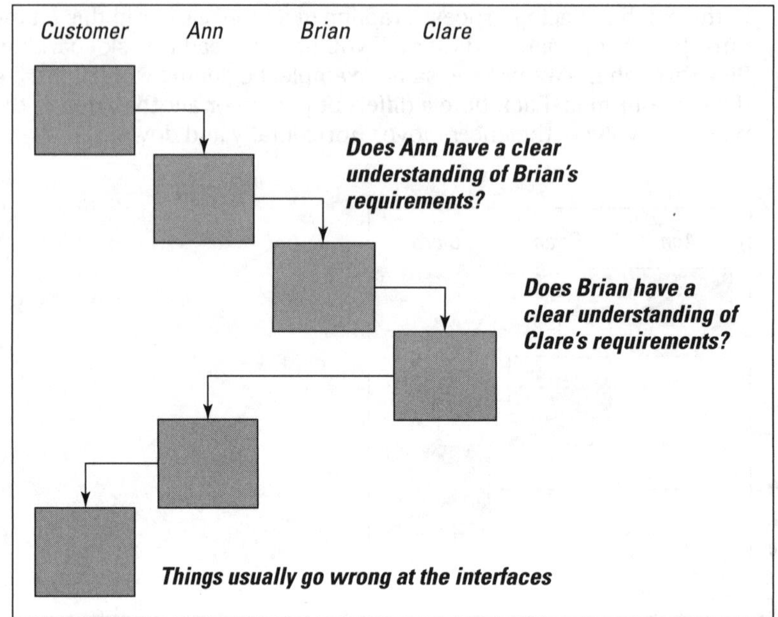

Figure 5-5: Highlighting the interfaces.

Measuring time can highlight other improvement opportunities, as shown in Figure 5-6. For example, you may ask how long each step takes and why.

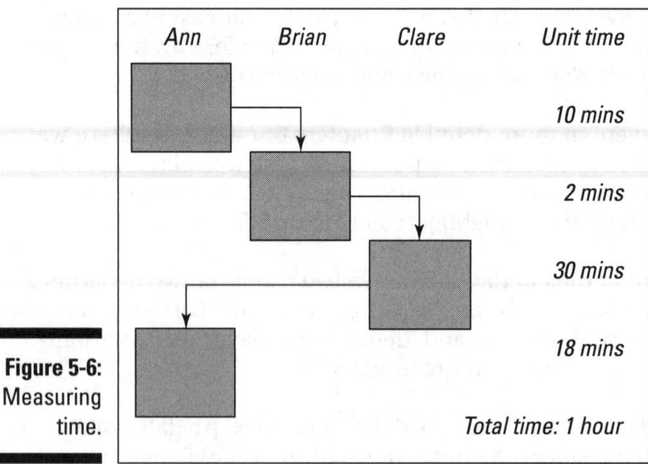

Figure 5-6: Measuring time.

In Figure 5-6 you're simply measuring *unit time* – the time it takes to complete this step. While this measurement could prompt some interesting questions, viewing the bigger picture is more helpful as it also includes the *elapsed* or *cycle time*. This measurement is the time it takes to complete the entire process, as shown in Figure 5-7. (Elapsed or cycle time is sometimes referred to as the *lead time*.)

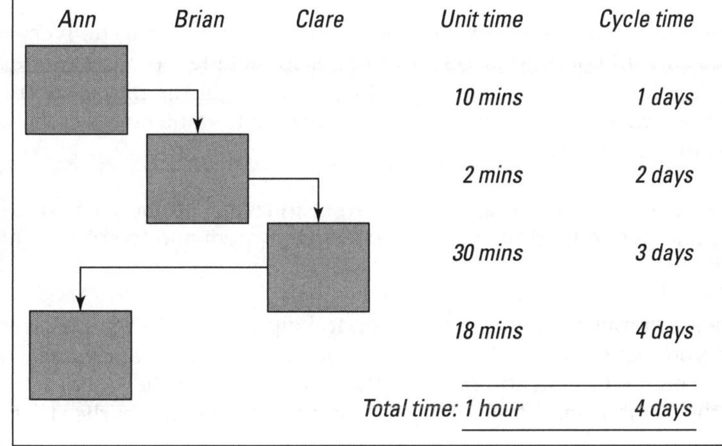

Figure 5-7: Cycle time showing dead time experienced by customers.

Building in the cycle time helps you identify bottlenecks and *dead time* – so-called because from the customer's perspective nothing's happening. In Figure 5-7, step number two is causing a bottleneck. If Brian's step can possibly be removed from the process, and Ann or Clare take up Brian's work, you may be able to halve the cycle time. Possibly Brian's step is a non-value-added step (explained in Chapter 9) and isn't needed at all.

Clear links are evident between measuring time and the theory of constraints, which we cover in Chapter 11. In your processes, try to identify and manage bottlenecks, to ask questions that clarify your understanding, and to always look for improvement opportunities. Earlier in this section, we listed some typical questions that your process picture may prompt.

Seeing the value in a value stream map

A value stream map is either an addition or an alternative to the deployment flowchart as a way of looking at how work gets done in your organisation. The term 'value stream' is a misleading description. The value stream map shows all actions, both value-creating and non-value-creating (more on value in Chapter 9), that take your product from concept to launch, and from order to delivery. These actions include steps to process information from the customer, and steps to transform the product on its way to the customer.

Toyota's Taiichi Ohno summarised the value stream nicely in 1978, when he said:

All we are doing is looking at a timeline from the moment the customer gives us an order to the point when we collect the cash. And we are reducing that timeline by removing the Non-Value-Added wastes.

Value stream maps follow a product's path from order to delivery to determine current conditions, but they can also include a picture of the actual working layout in the office or factory to highlight the impact of transport time, for example. You can create and use your value stream map in a way that works for you.

Ideally, your process map includes the external customer. You need to recognise and understand the whole process or system and to spot the moments of truth.

Process stapling is an ideal first step to help you create a value stream map – and you really do need to go the Gemba to see what's happening. For the lowdown on the Gemba and process stapling, check out the sidebar 'Genning up on the Gemba' and 'Practising process stapling' section earlier in this chapter.

The value stream map is similar to the format of a SIPOC diagram, which we talk about in Chapter 3. Ideally, your value stream map includes a picture of where the various activities happen and shows the flow of both materials and information, as shown in Figure 5-8.

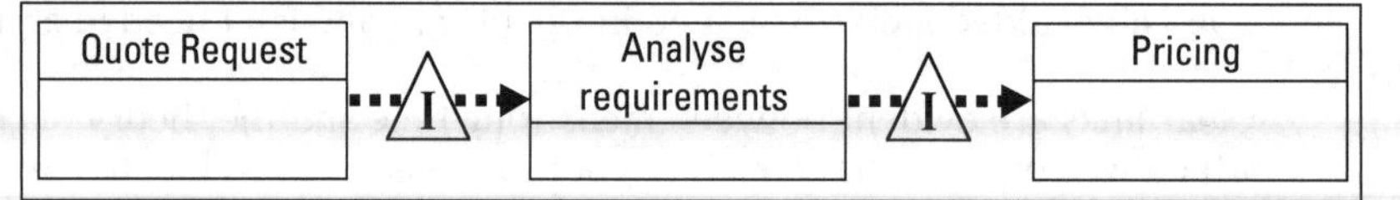

Figure 5-8: Part of a value stream map.

Figure 5-8 keeps things very simple, and looks very much like a SIPOC diagram. It includes some extra information; in this case, a triangle that identifies work in progress (the 'I' is for inventory) – work waiting to be actioned. In practice, value stream maps are straightforward, but they'll be a little more detailed than this example (see Figures 5-10 and 5-11) and will use more conventions than you use in a deployment flowchart (see earlier in this chapter). A selection of the more commonly used conventions is shown in Figure 5-9.

To draw your value stream map, work through the following steps:

1. **Identify the process you want to look at, agreeing the start and stop points.** Describing the product or service this process is supporting is also helpful.

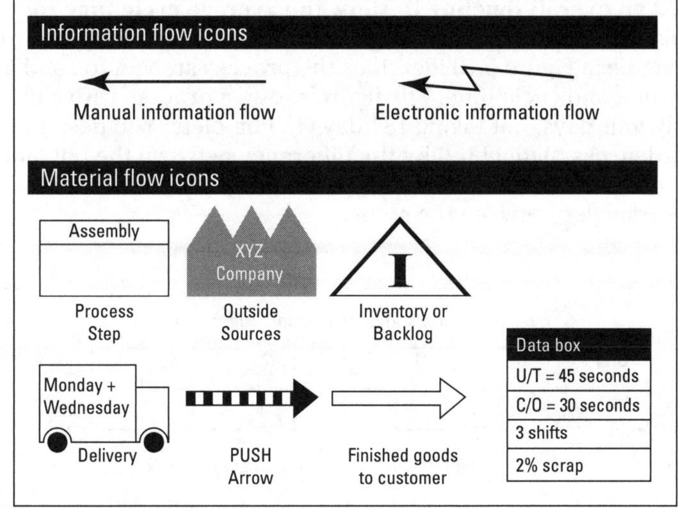

Figure 5-9: Value stream map conventions.

2. **Set up a small team to do the analysis.** The team should have knowledge of all the steps involved, from supplier input to external customer, so it must include people working in the process

3. **Go to the Gemba.** Go where the action is and watch what actually happens. Value stream mapping starts in the workplace.

4. **Working at a reasonably high level, draw a process map of the material/product flow in the whole value stream.** Some people prefer to do this exercise starting at the customer end and working backwards – rather like process stapling in reverse. Write down the steps as you go, rather than trying to remember everything. As well as material and product flow, remember to capture the information flow that causes product or material to move through the process.

5. **Identify the performance data you'd like to know.** Useful information often includes activity or unit time, cycle time, scrap or rework rates, the number of staff/resources, batch sizes, machine uptime, changeover time, working time, inventory, and backlog.

6. **Collect the data you need for each step in the process.** Add the data to your map in boxes. For example, in Figure 5-9, you can see a data box capturing a range of information, including unit time (U/T = 45 seconds).

7. **Add arrows to show information flows.** The value stream map shows information flow as well as material flow, separately identifying whether the information is sent manually or electronically (see the different symbols in Figure 5-9). The value stream map shows the information flow in the top half of the map, with the material flow below.

8. **Add an overall timeline to show the average cycle time for an item.** This timeline shows how long the item spends in the whole process. The example in Figure 5-10 identifies the process steps A to I and indicates the unit and cycle time. The figure shows a process with a unit time of only four days, but taking 187 days to complete! You need to look at the bottlenecks highlighted by the difference between the unit and cycle times, as well as the levels of work in progress or inventory identified in the triangles between the steps.

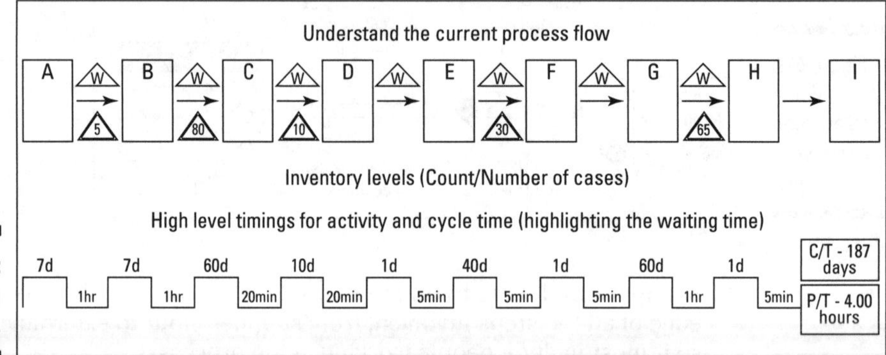

Figure 5-10: Identifying the delays.

Using averages is usually fine, but do recognise the danger of averages and remember that the actual times vary either side of the mean – known in the scary worlds of statistics and mathematics as 'variation'. We cover variation in detail in Chapter 8.

As an example of a value stream map, consider ABC Company's order process. The process begins with customer service receiving an email or telephone order. The product price is checked using the product price database.

Availability is checked in terms of stock inventory using the stock management system. If inventory cannot be allocated, the order is passed to the manufacturing team through the manufacturing order system and scheduled for production the next day.

The delivery date is determined, the customer is advised, and the order entry records are completed through the customer service order management system. The 'current state' picture of the value stream will resemble that shown in Figure 5-11.

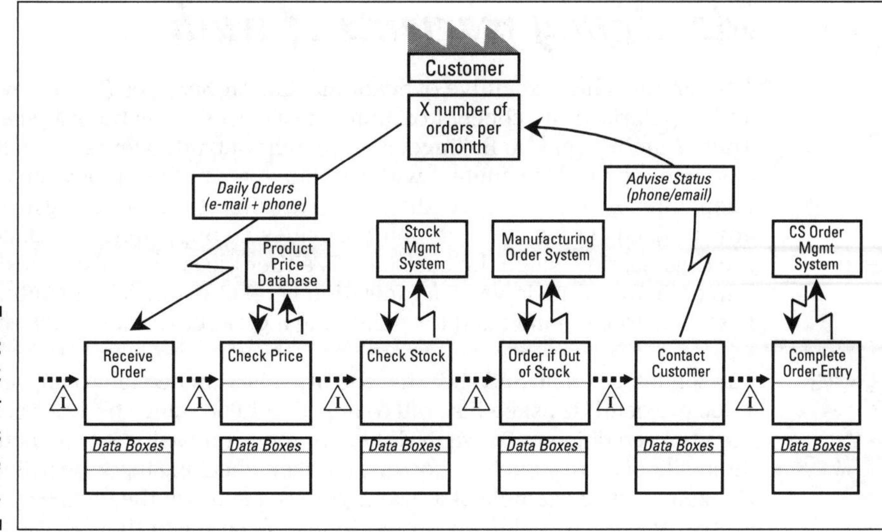

Figure 5-11: The ABC order process as a value stream map.

In the ABC example, the current state map includes some triangles containing the letter 'I'. These triangles are for the levels of inventory, or work in progress. When you create a value stream map for one of your processes, you need to remember that the map describes the current state of your organisation – a snapshot in time. Whether people in the organisation feel the inventory isn't usually that high or low isn't relevant; for whatever reason, the inventory is what it is *right now* in order to have a complete view of things. After you create the current state picture, you add some *data boxes*. These boxes capture a number of variables, such as activity time and cycle time (just as in the deployment flowchart, described earlier in this chapter), and *changeover time*. Changeover time is the time it takes to set up the equipment to move from processing one type of product to another, or to close one system and open another. A focus on reducing changeover time was one of the keys to success for Toyota in gaining market share over many of the western car manufacturers, where it was referred to as SMED – *single minute exchange of die* (die are the casts and moulds in the production system). In the spirit of continuous improvement, Toyota still look to reduce changeover times.

Toyota's Shigeo Shingo believed the company could make huge gains if changeovers could be actioned more quickly. He set a target to reduce any set-up time by 59/60ths. Shingo felt that many companies had policies designed to raise the skill level of their workers but few had implemented strategies to lower the skill level required by the set-up itself.

Changeovers and set-ups aren't relevant only to manufacturing companies and processes – they're just as relevant to service organisations.

Identifying moments of truth

Jan Carlzon, chief executive of Scandinavian Air Services (SAS), developed and popularised the concept of moment of truths in his book *Moments of Truth* (Cambridge, MA: Ballinger). A moment of truth occurs every time a customer comes into contact with a company, whether in person, on the telephone, by post, when reading company literature, or seeing a company advert. Each customer touch point provides an opportunity to make or break the organisation, since the customer is either pleased or displeased with the outcome. Everyone in your organisation is responsible for the outcome of customer touch points and for delivering a great customer experience.

Carlzon informed all SAS staff that the organisation needed to improve by 1,000 per cent! He asked his staff to improve 1,000 things by 1 per cent and then to keep doing it. He wanted them to focus on customer contacts – the moments of truth – such as booking a ticket, checking in, or boarding a plane. Carlzon used an example of a passenger pulling down the meal tray. If the tray was dirty, what would the customer think? What might that tell the customer about the maintenance of the plane?

To achieve what your customers want, you need to understand the many moments of truth opportunities that exist and find ways of enhancing the customer's experience. Process stapling, deployment flowcharts, and value stream maps can help you identify both internal and external customer touch points.

Part III
Assessing Performance

'I think there's a big change on the way, JB.'

In this part . . .

Here we look to see how well the work gets done. Are you meeting your customers' requirements in the most effective and efficient way? Managing by fact is a key principle in Lean Six Sigma, so having good data is vital. Data collection is a process in itself, and we present a five-step approach to ensuring you have an appropriate plan in place.

After you have your data, you need to decide how best to present and interpret it. We cover the importance of control charts to help you identify process variation so that you know when to take action and when not to.

We also look at developing an appropriately balanced set of measures to help you understand what influences and affects your results.

Chapter 6
Gathering Information

In This Chapter

- Understanding the difference between good and bad data
- Deciding how to collect your data
- Recognising that data collection is an ongoing process

Managing by fact is one of the key Lean Six Sigma principles. You need accurate, consistent, and valid data to manage in this way. This chapter focuses on developing a data collection process to ensure the data you collect meets these criteria.

You need to view data collection as a process that needs managing and improving just like all your other processes.

Managing by Fact

Whether you manage a day-to-day process or lead an improvement project, you need accurate data to help you make the right decisions. The following quote summarises the importance of fact:

> *Unless one can obtain facts and accurate data about the workplace, there can be no control or improvement. It is the task of the middle management and managers below them to ensure the accuracy of their data which enables the company to know the true facts.*
>
> – Kaoru Ishikawa, What is Total Quality Control? The Japanese Way

The following sections highlight the importance of good data, focus on the need to review your existing measures and develop an effective data collection plan to help you manage your processes.

Realising the importance of good data

Good data may prompt you to implement an improvement project by highlighting poor performance against the CTQs (see Chapter 3) or show you opportunities to tackle waste (see Chapter 9). It enables you to understand the current performance levels of a process and provides you with the means to benchmark that performance and prioritise improvement actions.

When you undertake an improvement project, you need to analyse the causes of the problem you're tackling – good data helps you quantify and verify those possible causes. In developing solutions to address root causes, you need good data to help you determine the most effective approach.

You're probably aware of the phrase 'rubbish in, rubbish out', which is often applied to data. You need to ensure you have good data going into your various management information reports and analyses. For that you must have a sound data collection plan, and we describe the key elements you'll need a little later in this chapter. First you need to consider what you're measuring.

Reviewing what you currently measure

Many organisations have data coming out of their ears! Unfortunately, that data isn't always the right data. Sometimes organisations measure things because they *can* measure them – but those things aren't necessarily the right things to be measured and the resulting data doesn't help you manage your business and its processes.

Sometimes data isn't accurate – intentionally or not – and even if the data is accurate, it may be presented in a way that makes interpretation difficult. Managers often present data as a page full of numbers to encourage comparisons with last week's results or even the results for this week last year. This situation is compounded further if the results show only averages or percentages and you can't understand the range of performance or the variation in your process performance. This range and variation in performance is what your customers will be experiencing (see Chapters 4 and 7).

In Chapter 7 we explain the importance of variation and how to use control charts to help you understand when and when not to take action. Without this understanding, a tendency exists to take inappropriate actions because managers are making the wrong decisions.

Figure 6-1 provides an ideal format to help you review your measures.

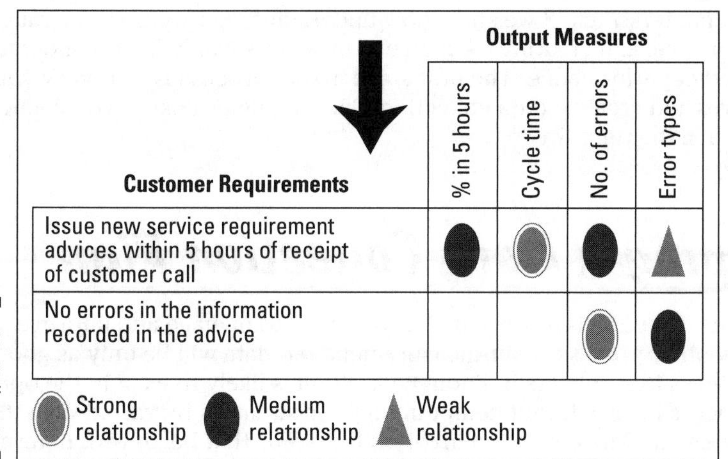

Figure 6-1: Getting the measure of the CTQs.

Deciding what to measure

You probably know that choosing what to measure and how to present your data are important. But so too is deciding what *not* to measure. Lean Six Sigma requires you to manage by fact and have good data – but that doesn't mean you need more data than you currently produce. It means you have the right data.

You need to review the data you currently have and decide whether it really is helping you manage your process. Does the data add value or is it a waste? Who uses the data? How and why is the data used? The CTQs provide the basis for your process measures and you need to consider whether your current measures help you understand your performance in meeting them. Figure 6-1 uses symbols to identify the strength of your performance measures in relation to the CTQs.

If you look at the CTQ for delivery within five hours, you can see that the first measure in the matrix, the percentage issued within five hours, is rated only as a 'medium strength'. That measure tells you how many cases are processed within the CTQ service standard, but it doesn't tell you anything about the actual results, and the range of performance. Some cases will have been actioned in one hour and some in ten hours, for example. This really important information is provided by the second measure, 'the cycle time', where you're recording the results of each and every case, or at least a representative sample. With this information, you can determine the average performance, the range of performance, and, of course, you can extract the 'percentage within five hours' information because you can see how many cases took five hours or less.

In Chapters 7 and 8 we show how understanding variation, the range of performance, and creating a balance of measures can help you understand and predict performance. The first stage in that process is to review your measures and create a data collection plan that helps ensure you collect the right data in the right way.

Developing a Data Collection Plan

Data collection is a process that you need to manage and improve, just like any other process. Your measurement and data will be only as good as the process that collects it. Enough variation is likely to exist in the operational process itself, without compounding the situation by variation in the measurement. Data collection involves five-steps that begin with determining the output measures for your processes:

1. **Agree the objectives and goals linking to the key outputs from your processes that seek to meet the CTQs.**
2. **Develop operational definitions and procedures that help ensure everyone is clear about what's being measured and why.**
3. **Agree ground rules to ensure that you collect valid and consistent data.**
4. **Collect the data.**
5. **Carry on collecting the data and identify ways to improve your approach.**

Beginning with output measures

We begin with the end in mind by considering the output measures. By agreeing on the end goals for data collection, and linking the data to your key outputs, everyone in the team understands why they're measuring what they're measuring. After the output measures have been agreed, you need to develop some additional measures to help you understand how the inputs to your process and the various activities in the process are influencing the output results. Chapter 8 covers this and looks at the importance of getting a balance of input, in-process, and output measures to help manage your process.

Agreeing on goals and outputs is usually straightforward if you've described the CTQ customer requirements in a clearly measurable way (we explain how to do so in Chapter 4).

Use our suggested symbols in Figure 6-1 to check whether you have an appropriate set of measures. You need at least one strong measure for each CTQ.

When you have a collection of output measures, use Figure 6-1 to review whether your output measures are appropriate – this may be particularly relevant if you've only recently determined the CTQs. After using Figure 6-1 in this way, you may consider abandoning some measures and creating other, more appropriate ones.

Cycle time (sometimes referred to as lead time) is the most important data. If you simply measure whether or not each item meets the service standard, you don't know the range of performance being delivered. For example, you may see that the organisation processes 80 per cent of orders within the service standard of five hours, but you may not be able to see that some orders take one hour, some take two or three hours, and the 20 per cent that fail take at least ten hours. With the cycle time data you can understand fully what happens.

In Figure 6-2 we show a process trying to meet the customer's requirements. The feedback from the customer and the process highlights a gap that you need to close – that is, you need an improvement action.

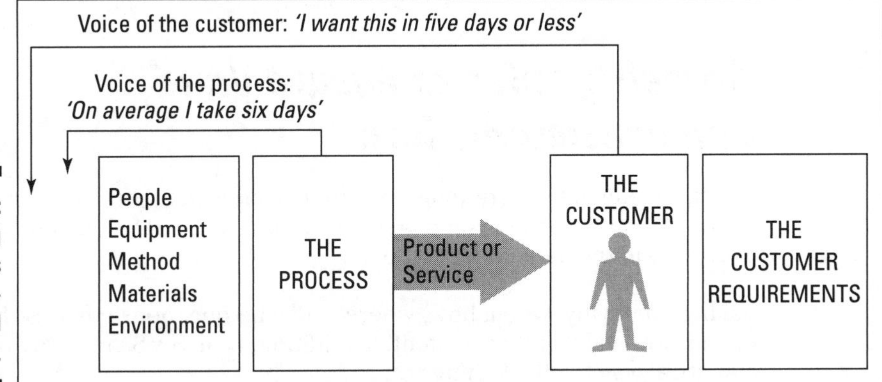

Figure 6-2: Matching the voices of the customer and the process.

In this example, we use average cycle times to represent the 'voice of the process'. Doing so isn't a good idea as average cycle times can be misleading. The average process performance in Figure 6-2 is six days, so the process doesn't meet the customer's requirement of 'five days or less'. But even if the *average* performance had been five days, the process probably wouldn't have been good enough. The customer sees every cycle time, not just the average.

Creating clear definitions

Describing your measures in a way that removes any ambiguity about what's being measured is the second step in your data collection plan. This description is called an *operational definition*.

When you know what you plan to measure, you need to provide clear, unambiguous operational definitions. These operational definitions help everyone in the team to understand the 'Who, What, Where, When, and How' of the measurement process, which in turn helps you produce consistent data. For example, if you measure cycle time, you define when the clock starts and finishes; which clock you use; whether you measure in seconds, minutes, or hours; and whether you round up or down.

The 1999 launch of NASA's Mars Lander is a famous example of murky definitions. This $125 million rocket was designed to investigate if water had existed on the red planet. Unfortunately, the rocket disappeared, never to be seen again. The cause was rather embarrassing: the team that built the spacecraft and managed its launch worked in feet and inches . . . but the team responsible for landing the craft on Mars worked in metric – and no one had thought to convert the data. As a result, the angle of entry into Mars was too sharp and the rocket burned up.

Agreeing rules to ensure valid and consistent data

Having an effective operational definition is important, but you also need to be able to validate the results. Asking yourself if the data looks sensible is the third step in the data collection plan.

Asking what, why, when, how, where, and who questions ensures that your data is both valid and consistent. In addition, Lean Six Sigma provides a statistical way to check things out.

Gauge R and R is a technique for assessing the repeatability and reproducibility of the measuring system – it confirms how much the measurement system contributes to process variation (see Chapter 7 for more on process variation).

Repeatability is a measure of the variation seen when one operator uses the same system, while *reproducibility* is a measure of the variation seen when different operators use the same system. To check repeatability, you ask someone to measure a batch of products and record their results, and then to measure the same batch again. If they don't get the same results, you need to decide whether the difference is important. We offer some broad guidelines to help you make that judgement in a moment.

To check reproducibility, you ask someone else to measure the same batch of products and see if their results are different. Again, if a difference does exist, you need to decide whether it's important and if action is needed to improve the measurement system.

In our example in Figure 6-3, two people – Timekeeper A and Timekeeper B – check the same batch of products in a random sequence. By averaging the difference of the two readings over the number of products in the batch, we can infer the gauge R and R.

Measure	Timekeeper A	Timekeeper B	Tolerance
Vet form	45	41	9.30%
Add info. to form	90	89	1.12%
Update records	175	177	1.14%
Print agreement	100	95	5.13%
Issue to customer	66	72	8.70%
Total Time	**476**	**474**	**0.42%**

Figure 6-3: Checking out the measurement system.

In Figure 6-3, gauge R and R is good for total time at 0.42 per cent, but is less accurate for the sub-processes. Overall these results are very good, but we could try to improve 'Vet form'.

Determining what's good in gauge R and R terms is somewhat subjective and there aren't any truly right answers. We can offer some broad guidelines but when you decide whether to take action, much depends on the process and the consequences of inaccurate data. Generally, if gauge R and R exceeds 10 per cent you should look to improve the measurement system, perhaps focusing on a better operational definition, for example, or using more accurate measuring equipment. If gauge R and R exceeds 25 per cent – change the measurement system!

When health and safety, regulatory, or important financial issues are involved, the gauge R and R guidelines need to be a lot tighter. You want very accurate and consistent data if you're making a decision that affects braking distances in the testing of a new car design, for example. Lives could be at stake if the gauge R and R is more than a fraction different. Measuring how long a telephone call takes in a call centre, in contrast, isn't so important.

Figure 6-3 covers *continuous data* – that which can be measured on a continuous scale, such as processing time. *Attribute data* includes whether or not something is present, or is right or wrong, and categories of items, such as types of compensation claim, complaint, and financial standing. To measure gauge R and R for attribute data, you ask a number of people in the process team to classify the items in a batch into the various categories. You can then compare their assessments both with one another and with an expert's assessments. Doing so ensures consistent classification by the process team and sometimes highlights training needs too.

In Figure 6-4, you can see how assessors A and B classify claims consistently between them, but aren't in line with the expert's assessment. This finding indicates a need to improve the quality of the training given to the assessors so that their classification is in line with the expert's view.

Claim Number	Expert's Classification	Assessor A	Assessor B
1	AA	AA	AA
2	AB	AA	AA
3	AA	AB	AB
4	AC	AC	AC
5	BB	BB	BB

Figure 6-4: Attribute data in action.

Collecting the data

Step four of the data collection process covers how you actually collect the data. You'll almost certainly collect some of it manually. Data collection sheets make the process straightforward and ensure consistency. A data collection sheet can be as simple as a check sheet that you use to record the number of times something occurs.

The check sheet is best completed in time sequence, as shown in Figure 6-5. This real example shows data from the new business team of an insurance company processing personal pension applications from individual clients. It captures the main reasons why applications can't be processed immediately; daily recording the number of times these different issues occur. On a daily basis, you can see the number of 'errors' and the number of application forms, and in Figure 6-5 we've recorded the proportion of errors to forms. Looking across the check sheet from left to right, you can see that we've recorded the total errors by type and have determined their percentage in relation to the whole. This check sheet links neatly to a Pareto analysis, which we show in Figure 6-6. Here, the 80:20 Pareto rule means that generally 80 per cent of the errors are caused by 20 per cent of the error types. Your analysis won't always result in precisely 80:20, and in our example, the main causes of the problem, C and E, account for almost 75 per cent of the errors.

The Pareto chart in Figure 6-6 highlights this fact. The cumulative percentage line helps you decide which errors to focus on. If you tackle type C errors, you'll address 39.4 per cent of the problem, but if you also address type E errors, you'll cover 73.9 per cent. You can tackle the smaller errors, A, B and D, later on.

Ref	Characteristic	M	T	W	T	F	M	T	W	T	F	Total	%
A	form not signed	2	1	0	1	0	1	1	0	0	1	7	8.3
B	no part number	1	0	2	1	1	2	0	0	0	0	7	8.3
C	address missing	5	2	3	2	3	4	3	5	3	3	33	39.4
D	no cheque	1	0	1	1	1	0	1	1	1	1	8	9.5
E	wrong amount	3	4	1	3	1	2	3	3	4	5	29	34.5
Total errors		**12**	**7**	**7**	**8**	**6**	**9**	**8**	**9**	**8**	**10**	**84**	**100**
Total forms		**24**	**20**	**21**	**18**	**18**	**24**	**16**	**20**	**14**	**22**	**197**	
Proportion		**.5**	**.35**	**.33**	**.44**	**.33**	**.37**	**.5**	**.45**	**.57**	**.45**	**.43**	

Figure 6-5: Checking out the check sheet.

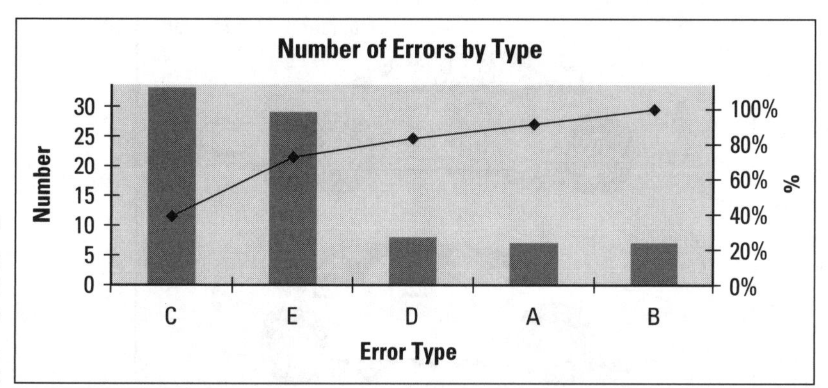

Figure 6-6: Looking at the vital few with Pareto.

Even if you use a computer system to automatically generate your data, design the data collection form on paper first. Doing so helps you think through all the details you may need, such as whether and to what extent to take account of segmentation factors (see Chapter 3), which could include different customer or product types, for example.

A *concentration diagram* is another form of data collection sheet. This technique is good for identifying damage to goods in transit, for example by recording where on the product or packaging dents and scratches occur. Car hire companies often ask customers to complete concentration diagrams. On a picture of the car, customers have to highlight existing damage, such as dents and scratches. Upon return of the vehicle, the leasing company then checks to see if any further damage has occurred. See Figure 6-7.

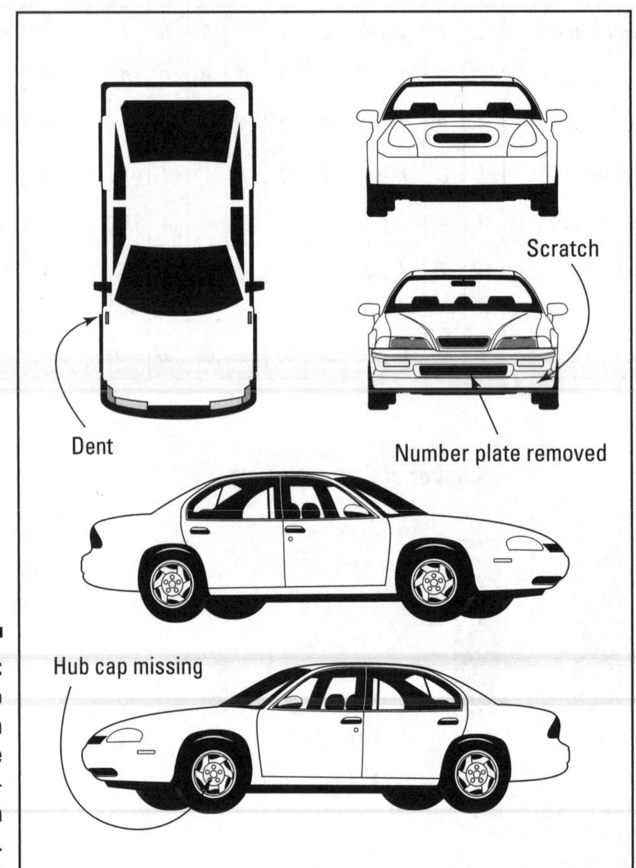

Figure 6-7: Coming up to scratch with the concentration diagram.

Identifying ways to improve your approach

The fifth step of the data collection process reminds you that data collection is a process and needs to be managed and improved just like any other. Even after you find your initial data, carry on collecting more to identify ways to improve your approach. You've determined the data showing your output performance. Now identify and measure the *upstream variables* that influence the output performance of your process. We cover upstream variables in detail in Chapter 8, but typical variables include volumes of work, supplier accuracy, supplier timeliness, available resources, and in-process cycle times. Measure these upstream variables on a daily basis.

Figure 6-8 provides a data collection summary. Use it to ensure you've covered all aspects of your data collection plan; doing so should lead you to collecting data that is accurate, consistent, and valid.

Data Collection summary

Type of measure	**What?** What are we measuring?	**Why?** Why are we measuring this?	**How?** How do we collect and record the data?	**When?** When do we collect the data?	**Where?** Where in the process?	**Who?** Who will collect it?
Output						
In-process						
Input						

Figure 6-8: Pulling the data collection plan together.

Enhance your summary by using icons to show how you present your data, for example images of Pareto diagrams, or control charts which show the variation in your performance (we describe these in detail in Chapter 7).

Chapter 7

Presenting Your Data

In This Chapter

- Investigating variation
- Using control charts
- Looking at different ways to display data

This chapter introduces the importance of understanding and identifying variation. If you can identify what type of variation you're seeing in your process results, you can determine whether action is needed or not, and avoid taking inappropriate action and wasting effort.

Control charts can be used to identify variation and this chapter covers how to use these powerful data displays. We focus on the most commonly used type, the X moving R, or individuals, control chart.

Later in the chapter we refer to some other data displays and ways to assess variation, looking at histograms and hypothesis tests.

Delving into Different Types of Variation

Things are seldom exactly the same, even if at first glance they appear to be so. Variation exists in people's heights, in the many shades of the colour green, in the number of words in each sentence of this book, and in the time different people take to read this book.

Variation comes in two types – common cause and special cause:

- **Common cause or natural variation** is just that – natural. You should expect it, you shouldn't be surprised by it, and you shouldn't react to individual examples of it.
- **Special cause variation** isn't what you expect to see – in the context of your processes, something unusual has happened that's influencing the results.

You can use statistical process control (SPC) and control charts to identify and define variation in your business processes and we explain just what these are and how to use them in the later section 'Recognising the Importance of Control Charts'.

Defining the type of variation is important as it ensures you take action only when you need to. Confusing one type of variation with the other creates problems.

Understanding natural variation

Natural variation is what you expect to see as a result of how you design and manage your processes. When a process exhibits only natural variation, it's in *statistical control*. Being in statistical control doesn't necessarily mean that the results from the process meet your customer CTQs (Critical To Quality elements of your offering – see Chapter 4) but it does mean that the results are stable and predictable. If the results don't meet your CTQs, you can improve the process using DMAIC (Define, Measure, Analyse, Improve, and Control – see Chapter 2).

To determine whether the variation is natural or special, try the following simple experiment with some colleagues.

First, write down the letter 'a' five times. This in itself forms the basis for an interesting discussion on giving clear instructions so that everyone understands the requirements. You may find that some people write their 'a's across the page, and others down the page. Some use capital letters, and others lower case. One or two may even write 'the letter "a" five times'!

Now look at your own letters and ask whether they're all the same. Each 'a' is probably slightly different, but generally they're likely to be pretty similar and at least each one can clearly be identified as a letter 'a'.

The difference between your letters is natural variation, and your process for producing the letters is stable and predictable. If you repeat the exercise, you're likely to see the same sort of variation. To reduce the variation, you need to improve the process, perhaps by automating your writing or introducing a template. We continue this exercise on page 99.

Spotlighting special cause variation

Special cause variation is the variation you don't expect. Something unusual is happening and affecting the results. Special cause variation may occur if you don't identify an important 'X variable', which influences your process results, or if you don't manage the variable appropriately. The Xs will include a range of factors, for example the accuracy and timeliness of the inputs to your process that you receive from suppliers, or the level of rework within your process (for more on X variables, see Chapter 8).

When a special cause exists, the process is neither stable nor predictable. You need to take action to identify the root cause of the special cause, and then either prevent the cause from occurring again or build the cause into the process.

Not all special causes are bad. Sometimes they provide evidence that an improvement has worked. We describe how you can identify special causes later in this chapter, but first we need to stress why doing so is so important.

Distinguishing between variation types

You need to be able to tell the difference between the two types of variation. If you think something is special cause variation when in fact it's natural, you may inadvertently tamper with the process and actually increase the amount of variation. Likewise, if you think something is natural variation when it's really special cause, you may miss or delay taking an opportunity to improve the process.

Avoiding tampering

In the 'Understanding natural variation' section earlier in this chapter, we ask you to write down the letter 'a' five times as an example of natural variation. We suggest that, to reduce the amount of variation, you need to review and improve the process. In this section we show what happens if you tamper with the process by reacting to an individual example of common cause variation.

As an example, imagine that your manager doesn't understand the importance of distinguishing between natural and special cause variation. She wanders through your work area to see the output being produced. She feels that your letter 'a's show too much variation and asks you to show how you produce the letters. As you begin to demonstrate, your manager asks you to stop writing and points out that using your other hand is much better – after all, this is the hand she uses!

If you try writing with your other hand, your results probably show increased variation, and chances are you take longer to produce the output. Now imagine the output goes through an optical scanner – depending on the quality of your letters when you write using your other hand, you might see further problems. Your manager then provides some unhelpful ideas to solve this problem, too.

Unfortunately, tampering happens all the time in many organisations. Managers often feel their role is to tamper.

Another example of tampering is pointless discussion. You may often see reports comprising pages of numbers that somebody expects you to understand and perhaps base decisions on. In Figure 7-1 we show a typical set of information that is practically meaningless to all but the person who created it.

Sales Performance - **May**										
	Location A					Location B				
PRODUCT	Previous month	Target	Current month	Target	% change from last year	Previous month	Target	Current month	Target	% change from last year
1	34	30	37	30	-5.4	59	50	56	55	-7.6
2	260	250	230	250	3.3	226	250	267	250	12.8
3	75	75	65	70	0.4	125	130	133	135	5.9
4	3	2	4	2	2.7	16	15	18	15	-6.7
5	4678	4750	4978	5000	10.6	1657	1600	1753	1700	5.9
6	930	950	1006	975	2.9	975	1000	952	1000	-1.5
7	950	975	1100	1050	-3.9		975	950	975	-6.2
8	43	45	48	45	-2.8	75	75	78	85	8.4

Figure 7-1: A typical data set providing little useful information.

Figures relating to sales activity often provide good examples of pointless data. You may hear statements such as, 'This week's figures were better than last week's, but not as good as the week before that' or 'It rained last Thursday, but the team did a great job this week' – almost certainly the differences in the weekly figures are a measure of the natural variation in the process and not due to a special cause.

Using control charts can help you make sense of the figures by enabling you to distinguish between natural and special variation – but you may need to change the way you think. The different thinking needed is described as you work your way through the data from Figure 7-1, eventually using it to create a control chart in Figure 7-3.

Displaying data differently

The data in Figure 7-1 don't tell you much. But if you present the data in a more visual form, you may begin to understand them. Figure 7-1 shows a typical set of row-by-column data, highlighting the sales performance for two different locations in the month of May. The figure refers to eight different products. You can see the number of actual sales, along with some targets.

Instead of giving the figures for only one month, a more useful method is to plot a graph, called a *run chart*, using figures for a series of months. A run chart plots the data in time order – a time series plot makes it easier to spot any trends. A run chart doesn't tell you whether the variation is natural or special – to know that, you use a control chart to see whether any changes are part of the natural variation of the process or whether they're unusual and need a second look.

In Figure 7-2 we use the figures for Location A and Product 3 to create a run chart that presents data through to the following March.

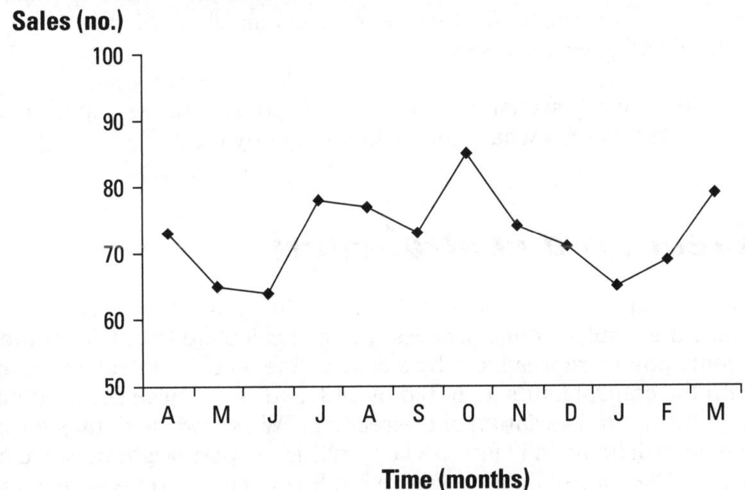

Figure 7-2: Presenting data as a run chart.

Recognising the Importance of Control Charts

Control charts provide the only way to identify and understand variation.

Walter Shewhart, who felt that businesses wasted too much time confusing the types of variation and taking inappropriate action, developed control charts in the 1920s. Shewhart envisaged the control chart as a way to simplify identification of variation. He knew his control chart should be a run chart showing the mean average, and also the upper and lower control limits (UCL and LCL). These upper and lower control limits show the natural range of the process results – but he was uncertain where to place these control limits.

Shewhart conducted thousands of experiments to determine the most appropriate position for the control limits. He discovered that the best positions were at plus and minus three standard deviations from the mean. We explain standard deviations in Chapter 1, but essentially, one standard deviation tells you the average difference between any one process result and the overall average of all the process results. It's a measure of variation and at plus one and minus one standard deviation from the mean average, you're likely to incorporate almost two-thirds of your total results. At plus and minus two standard deviations, you cover approximately 95 per cent of the results and setting the control limits at plus and minus three standard deviations includes 99.73 per cent of the data. Forget the statistics for the moment, though. Shewhart chose to place the control limits at these points, because here they work most effectively and economically to distinguish between natural and special cause variation.

Over time, many statisticians have reviewed Shewhart's experiments and concluded that Shewhart got his limits exactly right.

Creating a control chart

Control limits are calculated using the actual results from your processes. Using the results from a process, you can calculate the mean of the first twenty points, represented by a central line on the control chart, together with the control limits, denoted by UCL and LCL. These control limits represent the natural variation of the readings. We show the details for calculating the control limits in Figure 7-8 later in this chapter. Right now, we need to look at the control chart in Figure 7.4. If you feel apprehensive about calculations, don't worry too much about the maths: the calculations are relatively straightforward, and you can use software to do them for you.

Building on our example from Figures 7-1 and 7-2, we've built in some more data for Location A and Product 3. The control chart for the sales figures appears in Figure 7-3. The chart shows that the sales process exhibits variation, and that it is natural. We use the rules of statistical process control (SPC) to distinguish the type of variation. We cover SPC rules in the 'Unearthing unusual features' section later in this chapter, but for now work on the fact that, because all the data fall within the control limits, the readings reflect natural variation. This won't always be the case and you'll need to look out for unusual patterns in the data. These patterns are part of the rules we describe in the 'Unearthing unusual features' section.

If a process exhibits only natural variation, then it is in statistical control and is *stable*. Being stable means that the process results are predictable and you'll continue to get results that display variation within the control limits. Not reacting to individual data items is the key.

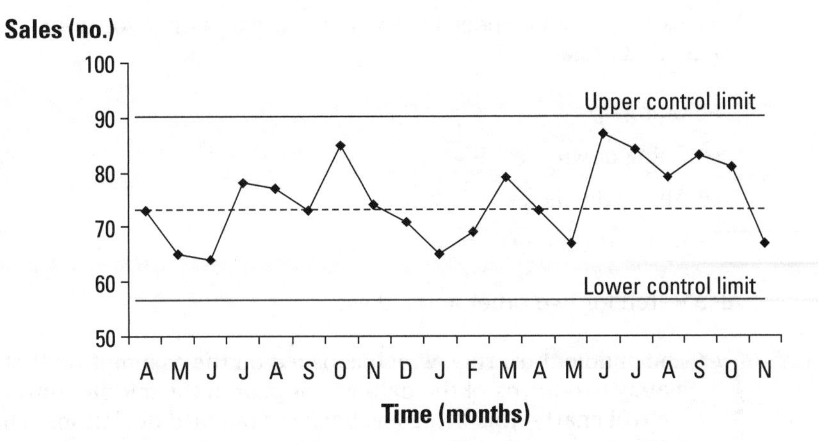

Figure 7-3: Control chart for a process exhibiting natural variation.

Just because all your readings reflect a process that's under control, stable, and predictable, doesn't mean your results are necessarily good. For example, you may find a large gap between the voice of the process and the voice of the customer (see Chapter 4 for more on these voices). Because the process is stable, you can at least review the whole process to find improvement opportunities.

When you take action to improve the process, you must update your control chart to show the changes. Charts should provide a 'live' record of what happens – a 'clean' control chart probably isn't being used properly.

Unearthing unusual features

You can identify special causes of variation in a number of ways. Noticing when a data item appears outside the control limits is an obvious one, as we show in Figure 7-4.

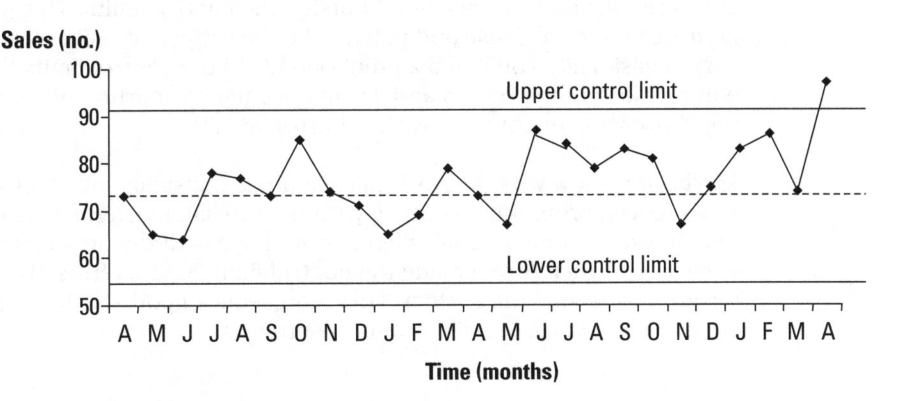

Figure 7-4: Occurrence of a special cause outside a control limit.

You also have some special causes to contend with if you spot a run of seven points that are all:

- Going up
- Going down
- Above the mean
- Below the mean

Also watch for two other anomalies:

- The middle third rule, which is based on the assumption that approximately two-thirds of the data will appear in the middle third of your control chart – this brings us back to standard deviations. The middle third of the control chart covers plus and minus one standard deviation, approximately two-thirds of your population data. If the spread of the data is out of line with this pattern, a special cause may be responsible. We aren't great fans of this rule because it can be applied too rigorously and lead to confusion – we tend to focus on the point outside the control limits and the run of seven rules.
- Unusual patterns or trends, where, for example, something cyclical is occurring or data is drifting upwards or downwards over time, but isn't by itself offending any of the other rules.

This chapter concentrates on the most important signal – a data point outside the upper or lower control limits. You need to find the root cause and then either prevent the special cause from occurring again (if the cause is bad) or build the special cause into the process (if the cause is good).

Statistical process control (SPC) is a large subject, and in this chapter we provide only the key points. For more detail, grab one of our companion editions – *Six Sigma For Dummies* or *The Six Sigma Workbook For Dummies*, both published by Wiley.

In Figure 7-4, you can see a point outside the control limits. This probably indicates a special cause and you need to investigate it, but be aware that very occasionally you'll find a point outside of the control limits that is a natural part of the process and lies in the small proportion of data outside of the 99.73 per cent covered by the control limits.

Maybe you know why the April sales figure is unusually high. Perhaps you ran a special promotion, coupled with the provision of a range of extra resources, resulting in a sales figure out of line with the previously expected values – and therefore outside the control limit. In SPC terms, the sales value for April is outside the system. This represents a good special cause – you need to see if you can build it into the process.

Sometimes you find a reason for an out-of-control signal that you can integrate into your improvement programme. As with this example, if you have a very high sales figure, and you know why, you can integrate this reason into the system and use it as part of an improvement strategy.

A special cause that most people are pleased to see is the proof that a change in the process has been successful. Figure 7-5 shows a situation where a process review has been carried out and an improvement action taken. The numbers on the vertical axis refer to the number of errors produced in sequential documents – perhaps the sales order forms.

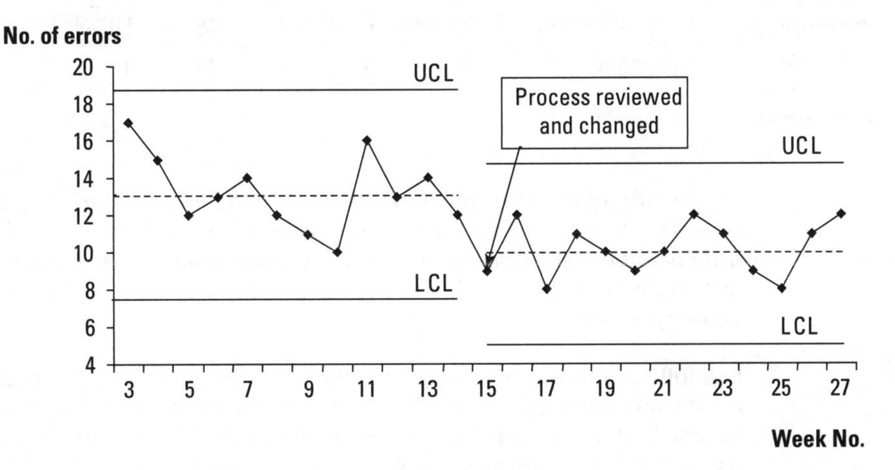

Figure 7-5: New control limits set after a process review and improvement action.

The results that follow the change to the process are all below the original mean (the dotted line), reflecting an improvement in the process. The control chart gives us evidence of a change for the better – in this case, seven consecutive points below the original mean. You can now recalculate the control limits and head down a track that may highlight new special causes to be actioned. Reducing variation is one of the key principles of Lean Six Sigma – and that's exactly what you're seeing in Figure 7-5.

Choosing the right control chart

You can use a number of different statistical process control (SPC) charts, but broadly they all follow the same concepts and rules. *Variable charts* measure time, volumes, or amounts of money, while *attribute charts* measure whether a particular characteristic is present, or right or wrong. Each type of chart has its own standard formula for calculating control limits, but generally the same rules apply to all.

Of the available control charts, the *X moving R*, or *individuals*, chart is the most versatile. This chart is ideal for measuring a range of things, such as cycle time performance and volumes, and to present attribute data such as proportions or percentages by treating them as individual readings. In the X moving R chart, the 'X' represents each of the data points recorded – perhaps a series of sales volumes, or the time taken to process each order. 'Moving R' describes the *moving range* – the difference between each consecutive pair of Xs, as shown in Figure 7-6.

Figure 7-6: Determining the moving range.

X	47	38	7	57	45	59
Moving R		9	31	50	12	14

Figure 7-7 shows the formula for the X moving R chart. The UCL_X and LCL_X represent the upper and lower control limits for the X data. The X with the little bar above it (X bar) is the mean average of all the Xs in the data you're using to construct your chart. R bar is the mean average of the moving range values you calculate – see Figure 7-6.

In addition to the control chart for the X values, you can also create a chart for the moving range values. The formulae make use of 'standard constant' values, in this case represented by A_2 D_3 and D_4. These have been calculated using statistics to provide shortcuts in the calculations.

The X moving R chart

$UCL_X = \bar{X} + (A_2\bar{R})$ $LCL_X = \bar{X} - (A_2\bar{R})$

A_2 is one of a number of constant values used in calculating control limits. If the moving range is determined by looking at each pair of Xs, then A_2 will always be 2.66.

The formula for the Moving Range part of the chart is:

$UCL_R = D_4\bar{R}$ $LCL_R = D_3\bar{R}$

D_4 is another of the constants used. Its value for the X moving R chart is 3.267. D_3 has no value for this chart, and the LCL_R will always be zero.

Figure 7-7: Looking at the formula for the X moving R chart.

Control charts are a key technique in the analysis, control, and improvement of processes. Be aware that top-level management should initiate control charts, not those working with the processes. Managers should understand variation and then demonstrate that understanding through their behaviour if their teams are to appreciate the benefits of control charts and SPC. This means that if the data is natural variation, managers shouldn't be tampering.

Examining the state of your processes

You can demonstrate your understanding of variation by using control charts to review performance in regular management meetings. Doing so can show you a number of important things that enable you to make more effective decisions. In particular, control charts help you to determine the state of your organisation's processes and potentially to transform the operational meetings. This would be the case where the main thrust of current meetings is asking why this week's results are worse than last week's, and using the data to blame people for poor performance. There's no understanding of natural or special cause variation and tampering is rife. This type of meeting and behaviour tends to lead to the distortion of data and process, and a failure to manage by fact.

A process can be in one of four states, which provide the basis for more effective discussion and action:

- In an **ideal state**, the process is in statistical control and meets the customer's requirements. If you use a traffic-light system at your operations meetings, you can think of this state as a 'green light' – you need no discussion about why this week's numbers differ from last week's. Knowing whether the process is in statistical control and meets the CTQs is what's important. By continuing to use the control chart, you can monitor the process and make sure it stays ideal. You may also want to improve on ideal – perhaps by delighting the customer or reducing the costs associated with the process – but to do that you need to implement an improvement project that looks at the whole process.
- A **threshold state** – or 'amber light' – describes a process that's in statistical control but doesn't meet the customer specifications. Your discussion then focuses on the action that you need to take to bring the process into an ideal state. Again, you don't need to discuss the variation between this week's and last week's numbers, because the process is predictable. Assuming a DMAIC improvement project is initiated (see Chapter 2), your ongoing discussions will concern the progress you make. By continuing to use the control chart, you can monitor the effectiveness of your improvement efforts and, in due course, provide evidence that they're working, perhaps with the identification of a special cause, similar to that in Figure 7-6.

- When a process is in the **on the brink state**, it meets the customer's requirements but is not in statistical control. The process has special causes and is unpredictable. This is a red light situation – at any moment, the process may slip into chaos.
- Chaos – the serious **red light state** – describes a process that's not in statistical control and doesn't meet the customer's specifications. By continuing to use the control chart, you can monitor the removal of special causes and the eventual improvement of the process. Removing these special causes from the process before you begin to change is important; otherwise, they'll impede your efforts to improve.

In a culture of continuous improvement, moving to an ideal state via a threshold performance is fine. Take a bite-sized approach to improvement, monitoring as you go.

As your use and understanding of control charts increases, you may wish to incorporate additional information concerning the capability of your processes. Capability indices help you understand more about how well your processes are doing in terms of meeting the CTQs.

Considering the capability of your processes

A process in statistical control is not necessarily a good process. The process is predictable, but it still may not meet your customers' CTQs. Two *capability indices* can be used to help assess your performance.

The capability indices compare the process performance and variation to the CTQs and provide both a theoretical and actual measure to demonstrate the relationship. They tell you precisely how capable the process is.

These indices are relevant only when your process is in control and the process is predictable. The first capability index – the *Cp index* – looks at the variation in the process compared with the specification of the CTQs.

Using the Cp index is like gauging whether or not you can get your car through a gap. Imagine that the width of the control limits in your control chart is represented by the width of the car in Figure 7-8, and the arch represents the width of your customer's specification limits, their CTQ. Consider whether you can drive through the arch.

Is the process 'Inherently Capable'?

The width of car represents the control limits for 'the process'

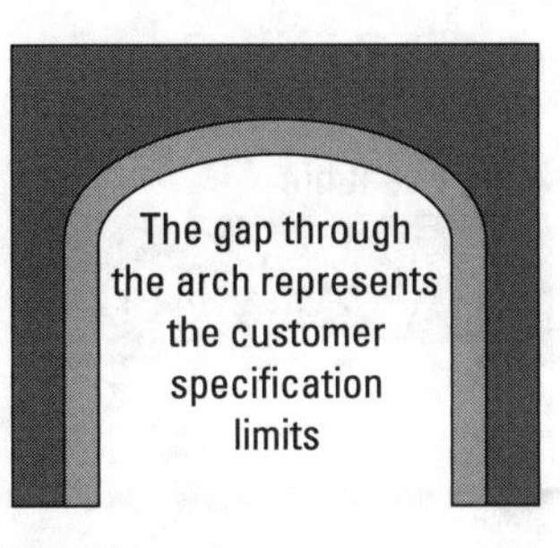

The process is 'Inherently Capable' but only if it's 'centred'

Figure 7-8: Taking your driving theory test.

In Figure 7-8, you can see that driving through the gap is possible – but only just. You need to drive very carefully and 'centre' the car. In other words, you need to line up the mean of the control chart with the mid-point of the customer's specification. The Cp index tells us how many times the car can, in theory, fit inside the arch. In our example, the car fits inside the arch just once, so the Cp value is 1.0.

Cpk is the second capability index and it describes the 'location'. Cpk tells you how well you're 'driving' – that is, how well you manage your process. The location describes the position of your process performance as presented by your control limits when compared to the CTQ specification.

In Figure 7-9, the driving needs some improvement. Here, the Cpk value is less than the Cp value. If Cpk is less than 1.0, it doesn't meet the CTQ. In process terms, you need to shift the mean by improving this threshold process. Reducing the variation also makes the 'fit' of the car in the arch that little bit easier.

Figure 7-9: The driving needs improvement.

To fully assess a process you need both the Cp and the Cpk values. You also need to be in control of the process for these indices to be meaningful. The Cpk value is never greater than the Cp value. When the process mean is running on the 'nominal' – the mid-point of the customer specification – Cpk and Cp are the same value.

Improving your processes requires you to increase the Cp value, making sure that it remains equal to Cp as far as possible.

Figure 7-10 shows the formula for calculating the Cp and Cpk values for the X moving R chart (see the 'Choosing the right control chart' section earlier in this section). The difference between the control limits, the UCL and the LCL, on your control chart covers six standard deviations. The difference between the upper and lower customer specification (USL – LSL) is usually referred to as the *tolerance*. The USL and LSL represent the range of the customer's CTQ. For example, they may want a product delivered within five days, but describe their requirement as an upper specification of five days and a lower specification of one day. Three days would be the nominal value of this specification, the mid-point, and the tolerance, the difference between the specification limits, would be four days.

Dividing the tolerance by the distance between your control limits (six standard deviations), you have the Cp value – the theoretical number of times you can fit the car within the arch. Your Cp value must be at least one if you are to meet the CTQ.

Figure 7-10 describes how the formula for working out your Cpk depends on the position of the mean on your control chart.

The Cp Index

$$Cp = \frac{USL - LSL}{UCL - LCL} = \frac{\text{Tolerance}}{\text{6 standard deviations}}$$

- For the process to be inherently capable, the Cp index needs to be at least 1.0.
- It may be inherently capable, but how well is it located? You need to use the other capability index, Cpk to find out precisely.

You need to use one or the other of the following formulae: Which one depends on the position of the mean on the control chart.

- If the mean is closer to the customer's upper specification limit, you use:

$$Cpk = \frac{USL - \bar{X}}{\text{3 standard deviations}}$$

- If the mean is closer to the lower specification limit, you use:

$$Cpk = \frac{\bar{X} - LSL}{\text{3 standard deviations}}$$

- 3 standard deviations is the distance from the mean of the control chart to the upper control limit; 6 standard deviations is the distance between the upper and lower control limits.

Figure 7-10: The capability formula.

In Figure 7-11, the process has a Cp value of 2.0 – in theory, it can fit inside the arch twice. If the Cpk is also 2.0, the process is very capable of meeting the customer specification and does fit inside the arch twice. On the other hand, if your driving isn't so good – that is, you don't manage your process well enough – you can see the effect as the Cpk value reduces, moving from left to right in the figure. When Cpk is below 1.0, you're unable to consistently meet the customer's specification.

The capability indices help give you a complete picture of performance. They can help you prioritise improvement action, too. By comparing the Cp and Cpk values of different processes, you can decide where to focus your improvement efforts, perhaps concentrating on those processes where the values are less than one.

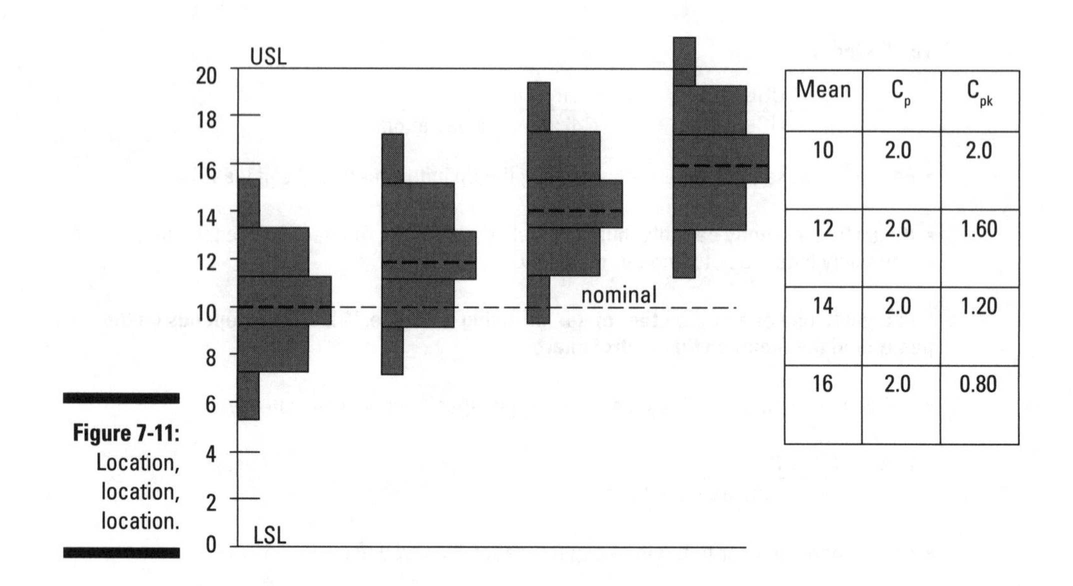

Mean	C_p	C_{pk}
10	2.0	2.0
12	2.0	1.60
14	2.0	1.20
16	2.0	0.80

Figure 7-11: Location, location, location.

Additional ways to present and analyse your data

As we explain in this chapter, process data may be presented in control charts, enabling you to determine the state of the process and its capability to meet CTQs. This information provides you with a clear picture of the action needed, if any, but other data displays can also help you analyse your data, including the histogram, for example, shown in Figure 7-12.

Histograms can be helpful in providing a picture of the mean and range of performance, and indeed the distribution of the data; they don't help you determine the *type* of variation that you are seeing, however. In Figure 7-12, the data are distributed normally, but this isn't always the case. Sometimes you might see examples of skewed data, where, perhaps, a lot of items are processed quickly but a long tail of data reflects items that are delayed for some reason – see Figure 7-13. These delayed items may be creating customer complaints or increasing your processing costs in some way, or both!

You need to understand the reasons for the delay, perhaps using a check sheet and Pareto diagram to present your results (see Chapter 6 for more on these). The histogram can also help you identify the need to segment your data, as shown in Figure 7-14.

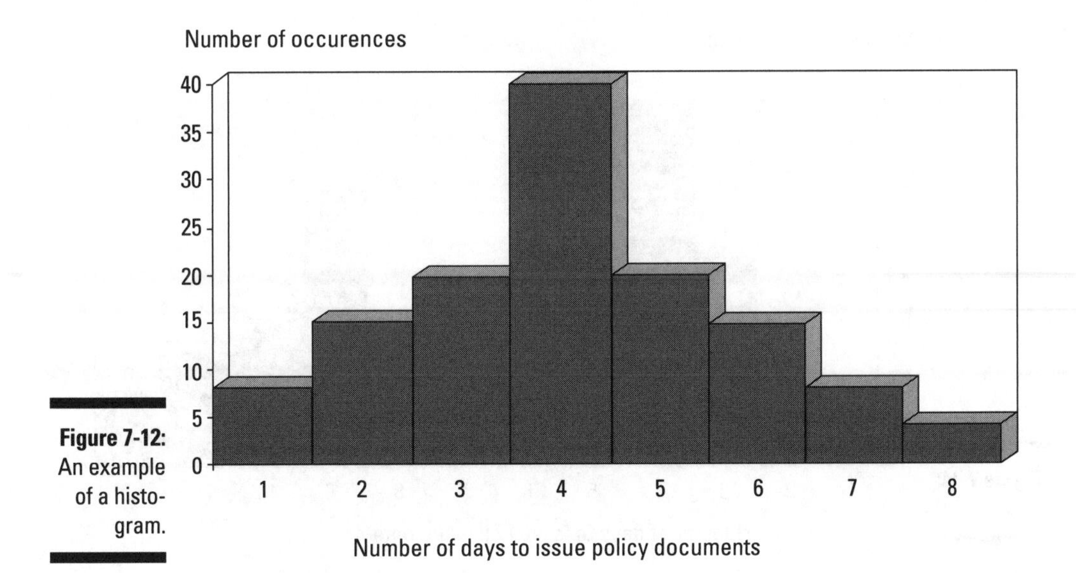

Figure 7-12: An example of a histogram.

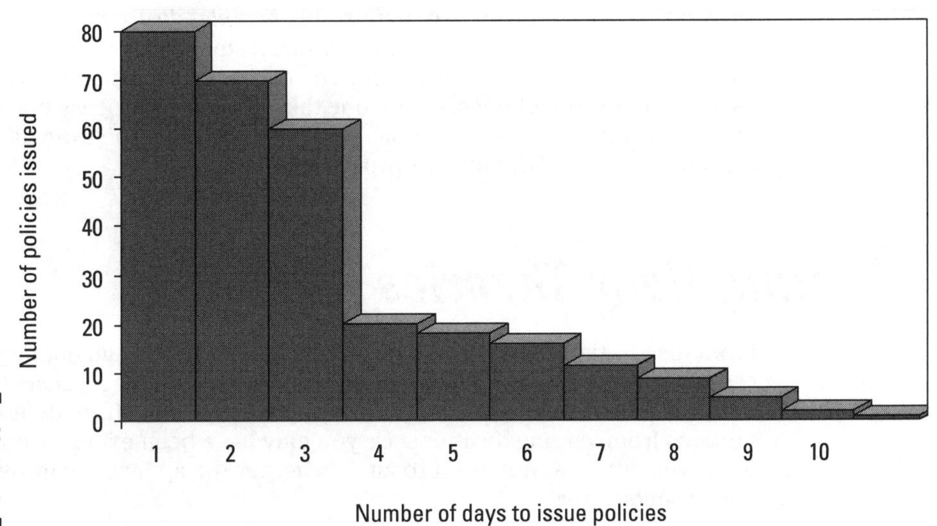

Figure 7-13: Looking at a long tail.

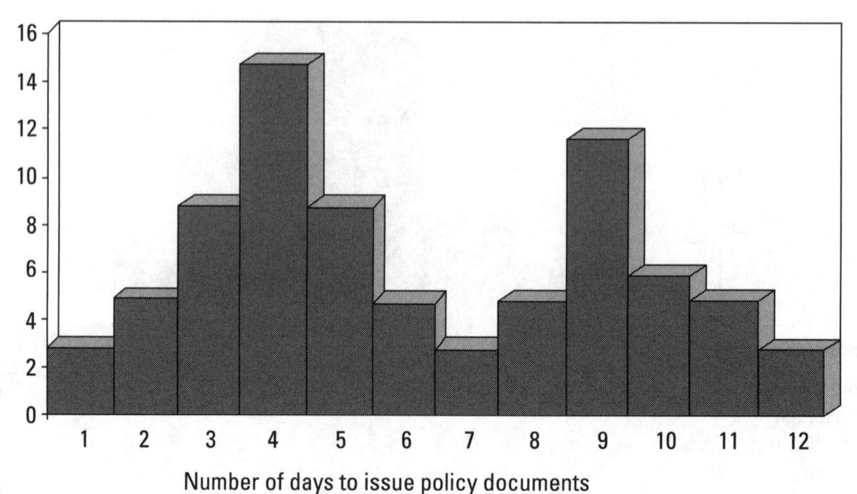

Figure 7-14: Twin peaks.

Looking like a camel with two humps, this *bimodal distribution* (two peaks) contains two 'populations', and to fully understand what the results are showing you need to separate them. The two populations could well be two different product lines, where one takes longer to process due to its increased complexity, for example, or the results might be from different locations dealing with different processes.

Testing Your Theories

From time to time, and particularly where you've been segmenting your process data, you need to know whether a statistical difference exists between data sets. The data might show, for example, the results from different teams, perhaps from varying locations. Or you may have been experimenting with improvement ideas and want to know whether the apparent improvement in your results is real.

You may be able to see a difference by viewing the shape of the data in a histogram, for example, or by comparing the mean or standard deviation, or indeed, the amount of variation. Even if you appear to see a difference, you may need to know how different your difference is. You may want to determine whether the differences are 'real' or just natural variation.

So, you need to decide whether the data has come from the same population or a different population. Possibly, the sample represents the future population to come. You need to determine if the data is really different – fortunately, you can use a hypothesis test to help you find out.

Hypothesis testing helps you find out if a statistically significant difference exists or not. This chapter provides only a brief overview of the tests, which are well supported by software programs such as Minitab or JMP, for example. Check out the Minitab or JMP websites or more details about hypothesis testing.

Creating two hypotheses for your tests, the null hypothesis and the alternate hypothesis, is the first step. The *null hypothesis*, usually expressed as H_0, proposes that no difference exists in the data. The *alternate hypothesis*, H_A, states a difference is evident. The alternative hypothesis is sometimes presented as H_1.

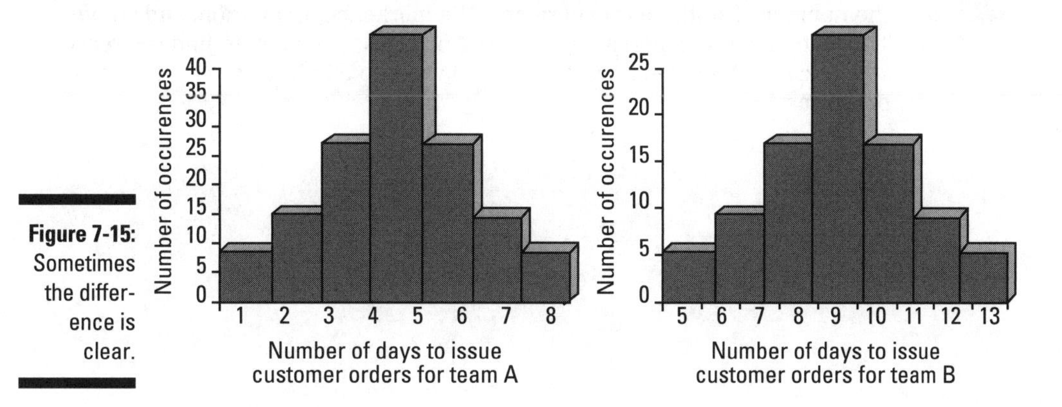

Figure 7-15: Sometimes the difference is clear.

Hypothesis tests are different from control charts; they don't look at ongoing data, but rather take a sample at a point in time. Usually, a 95 per cent confidence level is used, that is, you can be 95 per cent confident that the results display either a statistical difference or they don't.

Two hypothesis tests may be especially useful: the T-test and the ANOVA. The *T-test* looks at two sets of data, and the *ANOVA* considers three or more sets of data.

An example of using a T-test is determining whether a process change has really improved performance: you look at the before and after results, perhaps following a DMAIC project. An example of using ANOVA is comparing the results from several teams in order to identify whether one team is performing better than others, which perhaps provides an example of best practice to follow.

A *p value* determines whether a real difference exists in your data. Using the usual confidence level of 95 per cent, if p is less than 5 per cent (p = < 0.05), you can be 95 per cent confident that a difference exists – a 5 per cent chance of spuriously seeing a difference when one isn't there still exists, but the odds are overwhelmingly (19 to 1) against this being the case. If p is equal to or greater than 5 per cent (p > or = 0.05), you can conclude that insufficient evidence exists to reject the null hypothesis. You can interpret this conclusion in one of three ways:

- The samples come from the same original population.
- You have too much variation.
- Your samples are too small to detect any real difference.

Where a difference is evident, in the performance of teams at different branches, for example, don't jump to conclusions about why. The difference could be related to the way the data is collected, the size of the branch, the number of staff, their experience, the market segmentation, and so on. Through discussion and analysis of the process, you need to find the reasons, so you can build in best practice or find ways to eliminate the root causes of problems.

Chapter 8

Analysing What's Affecting Performance

In This Chapter

- Finding out what's at fault
- Using data to prove your point
- Introducing the maths of Lean Six Sigma

Whether you manage a day-to-day operation or are involved in a DMAIC (Define, Measure, Analyse, Improve, and Control) improvement project, you need to understand what factors can affect performance, especially if you encounter problems in meeting your customers' requirements. In this chapter, we introduce a selection of tools and techniques to help you identify the 'guilty parties'. We focus on how and how well the work gets done – the process and the data.

Unearthing the Usual Suspects

If you've seen the Oscar-winning film *The Usual Suspects*, you may remember its false trails and red herrings. Not until the closing scenes do you find out just who the guilty person is. Many of the usual suspects are innocent – just like in real life and in your search for the root causes of problems in your process.

People often jump to conclusions about the possible causes of problems. In many organisations, managers seem to 'know' for sure what the causes are; usually, however, a whole range of suspects influence performance and affect your ability to meet customers' CTQs (Critical to Quality) – but chances are only a vital few are actually 'guilty'. Consider the following fishy tale.

A shortage of oysters off the eastern seaboard of the US has occurred in recent years. Clearly, this is a symptom of global warming or pollution. But perhaps not: research carried out by a Canadian university shows how the functional elimination of large sharks from the east coast, especially the scalloped hammerhead, has inadvertently resulted in dwindling supplies of shellfish due to an increased population of rays who eat them – see Figure 8-1.

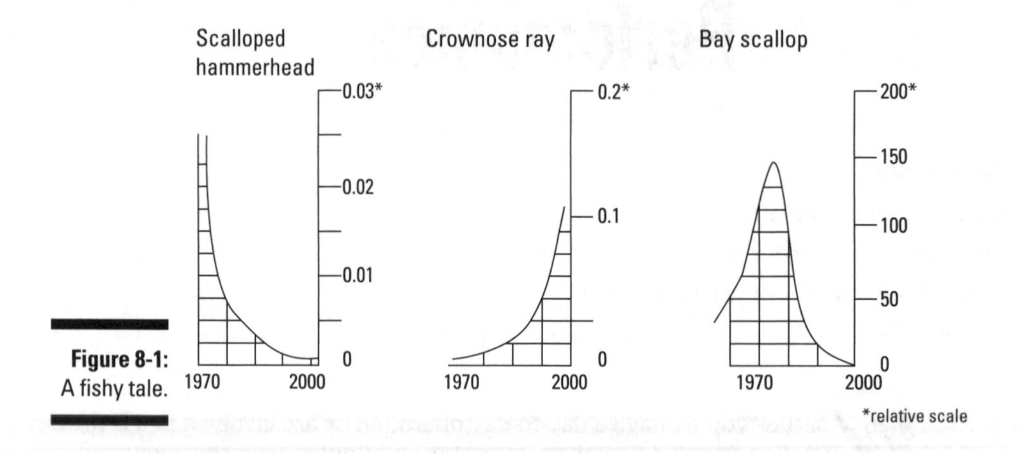

Figure 8-1: A fishy tale.

Generating your list of suspects

To find the guilty party, you generate a list of possible causes, check out each possible cause, and gradually narrow down the list. In this section we look at the methods available to help you root out the suspects.

Creating a cause and effect diagram

The fishbone, or cause and effect, diagram (see Figure 8-2) provides a useful way of grouping and presenting ideas from a brainstorm. Write your ideas on sticky notes so that you can move them around easily during the subsequent sorting process.

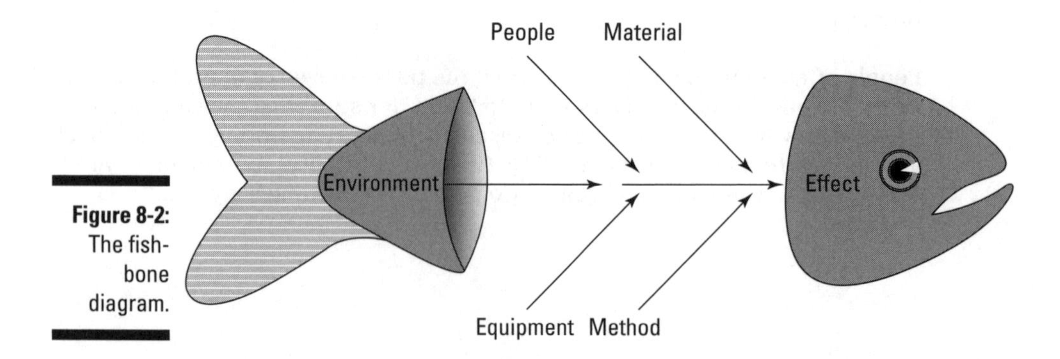

Figure 8-2: The fishbone diagram.

The head of the fish contains the question (make sure you choose a focused question or you'll end up with a whalebone!). For example, you might ask, 'What are the possible causes of delays in delivering customer orders?' or 'Why are there so many errors in our invoices?'

You can group the possible causes under whatever headings you choose. In Figure 8-2 we use the traditional headings of People, Equipment, Method, Materials, and Environment. You may find these headings useful in prompting ideas during the brainstorm, but be aware that they can also inhibit more lateral thinking.

Place your major cause headings on the left-hand side of the diagram, forming the main 'bones' of the fish. The brainstormed ideas (the potential main causes) form the smaller bones. For each possible cause, ask the question 'Why do we think this a possible cause?' and list the responses as smaller bones coming off the main causes. You may have to ask 'why?' several times to identify the real reason.

Use an interrelationship diagram next, to help you focus on the right Xs, as shown in Figure 8-3. Remember, the numbers next to the boxes represent arrows out over arrows in.

Instigating an interrelationship diagram

Using an interrelationship diagram (ID) helps you identify the key drivers behind the effect you're investigating in your fishbone diagram. We cover the ID in Chapter 2 where we show how you can use it with an affinity diagram – a really useful way of helping you get to the root cause.

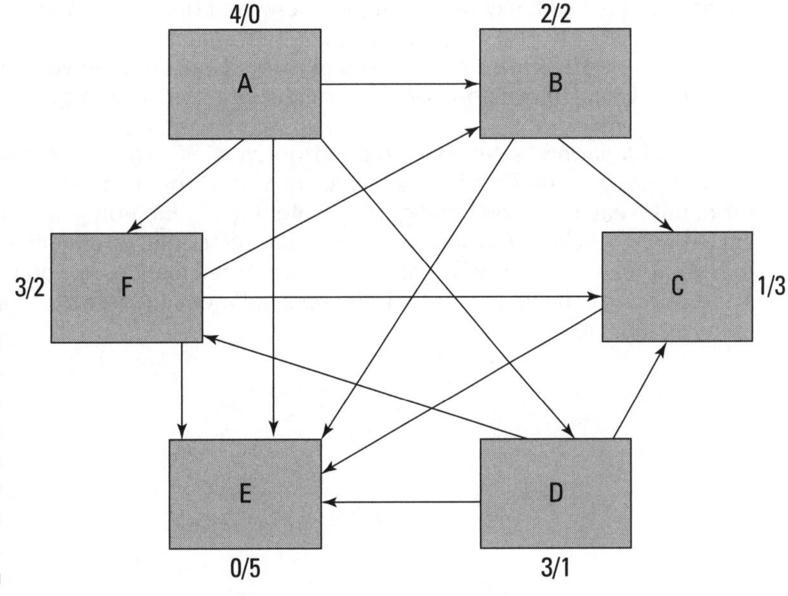

Figure 8-3: The fishbone diagram meets the interrelationship diagram.

Investigating the suspects and getting the facts

Managing by fact is vital, so validating the possible causes highlighted by your interrelationship diagram (see previous section) is the next step. All those possible causes are innocent until proven guilty. To validate your causes, you may need to observe the process and go to the Gemba (the place where the work gets done – see Chapter 2), or check out the data to see whether they confirm your suspicions. You'll probably need to collect some additional data to do this. Chapter 4 covers the development of measurable CTQs, which provide the basis for the measurement set of your process, and Chapters 6 and 7 introduce the importance of a data collection process, beginning with the need to measure the outputs of your process.

In Lean Six Sigma speak, the output measures are Y data, and the results here are influenced by the upstream X variables. Xs and Ys are actually just cause and effect. Individually and collectively, the various Xs influence your performance in meeting the customer CTQs, the Y variables. Sometimes, Xs are referred to as 'independent variables' and Ys as 'dependent variables'. Clearly, the Y results depend on you managing the Xs very carefully.

A SIPOC diagram (see Chapter 3 for the details) provides an ideal framework to help you think about all your process measures and now you need to pull together a set of X measures, if you don't already have them. A range of X variables will be coming into your process – the 'input variables'. These input variables affect the performance of the Ys, and may include the volume of activities, for example the number and type of new orders. These input variables may well concern the performance of your suppliers, too, perhaps in terms of the level of accuracy, completeness, and timeliness of the various items being sent to you. The inputs might be from customers or suppliers, but either way, they'll impact on how you perform. How often do you need to go back for missing information or clearer instructions, for example?

A range of X variables will exist in the process itself – the 'in-process variables'. Here, your deployment flowchart or value stream map (see Chapter 5 for details) can help you highlight the potential Xs, including activity and cycle times, levels of rework, the availability of people, or machine downtime, for example. Again, these Xs will affect your performance. As you identify the X measures you need, so you're building a balance of measures to help you manage your process.

Getting a Balance of Measures

To fully understand the performance of your process, you need a balance of input, in-process, and output measures, as shown in Figure 8-4, with perhaps one to three measures for each. You also need to recognise that the input and in-process variables will influence the results in your output variables, so your measures should clearly link together.

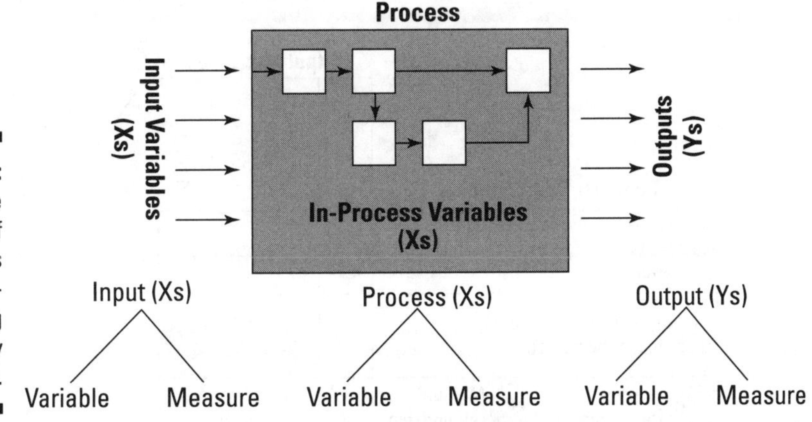

Figure 8-4: Getting the balance of measures and understanding how they interrelate.

Connecting things up

Figure 8-5 provides a reminder of how the CTQs are pulled together and incorporates Figure 6-1 from Chapter 6. This shows how you need to put the measurable CTQs into a matrix, determine the output measures, and assess the output measures as strong, medium, or weak.

Now you put the output measures into the Y to X matrix as shown in Figure 8-5, identifying the X variables and the corresponding measures for them. You also need to assess how these X measures link to the outputs, ensuring you have at least one strong X measure linking to each output measure, the Ys.

Ideally, you can determine precisely how the variables interrelate, considering the X measures as 'leading indicators' and the Y measures as 'lagging indicators' – that is, a clear cause and effect correlation is in place. To do this, you need to use a scatter diagram to help prove your point.

Voice of the customer	Key issues	Measurable CTQ
To meet timescales we need to know about new service requirements asap	Speed is key as the service must be completed within agreed timeframes	Issue new service advices within 5 hours of receipt
We must have the right details and information	Accuracy is vital to avoid wasted activity and lost time	No errors in the documentation

The Fishbone may have highlighted some gaps.

Do we have the right measures and an appropriate balance? Perhaps we need to segment the data in some way.

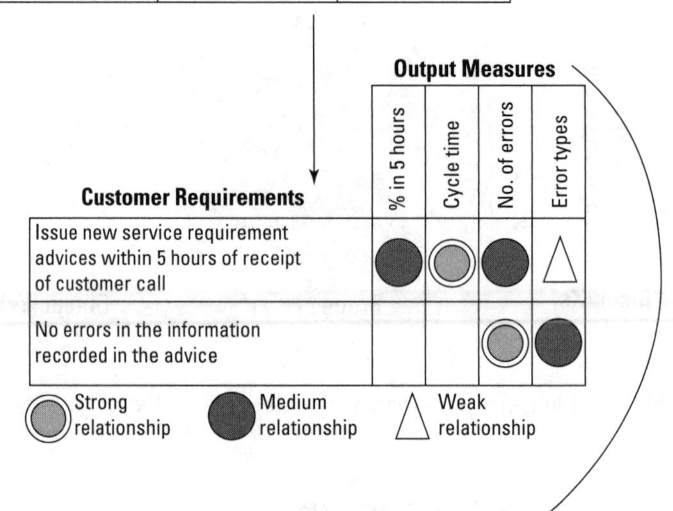

What are the input & in-process X variables?

What are the input & in-process measures?

Output Measures						
% in five hours						
Cycle time						
Number of errors						
Type of errors						

Strong relationship | Medium relationship | Weak relationship

Figure 8-5: Linking things up: Developing and reviewing the measures.

Proving your point

When you think you know the cause of the problem in your process, you may need to provide some evidence to back you up. For example, your boss may think they know the answer, but you may find something different due to your careful analysis of the facts.

You can use a simple matrix to show how the various snippets of evidence match against the suspects. This matrix is sometimes referred to as *logical cause testing*, where you summarise the possible causes of the problem, and show whether the various evidence you've gathered from your process and data analysis logically matches the suspects.

Using a *scatter diagram* can help you strengthen your case. A scatter diagram helps you identify whether a relationship exists between two variables and enables you to give a value to that relationship. The variables are the cause and effect – X and Y. You can use this method to verify potential root causes of a problem, or, for example, to validate the relationship between your input and in-process measures against your output measures. If your suspected cause (X) is real, then any changes in X produce a change in the effect (Y).

The dependent Y variable is always plotted vertically; the independent X variable is plotted on the horizontal axis. The data is plotted in pairs, so when X = 'this value', Y = 'that value'. We show four such pairs in the first example in Figure 8-6. In this example, a relationship seems to exist between speed and error rate – the faster we do it, the more errors we get. This correlation is positive because the values of Y increase as the values of X increase.

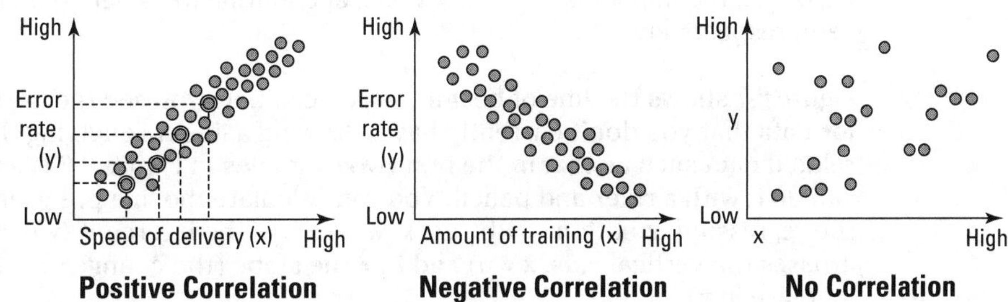

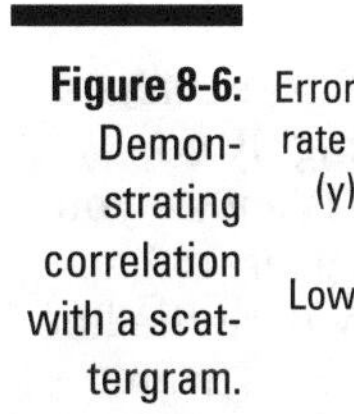

Figure 8-6: Demonstrating correlation with a scattergram.

The second example in Figure 8-6 shows a negative correlation – the values of Y decrease as the values of X increase, confirming our theory that investment in training leads to reduced error rates.

In the third example in Figure 8-6, no correlation exists, so our theory doesn't hold. Whatever the X is, it doesn't influence the Y results; do make sure, though, that the data has been segmented, otherwise a pattern might be hidden from view. Chapter 3 covers segmentation.

Seeing the point

Simply seeing the picture may be enough to demonstrate that you have or haven't found the root cause of your problem, but to strengthen your case you can put a value on factors by calculating the *correlation coefficient*, or r value. This value quantifies the relationship between the X and the Y – it tells you the strength of the relationship in terms of the amount of variation the X is causing in the Y results.

In a perfectly positive correlation, $r = +1$. In a perfectly negative correlation, $r = -1$. Usually the number is less than one, as the possibility of only one X affecting the performance of the Y is unlikely; generally, several will be evident. Almost certainly, however, one X will be creating the main effect.

The correlation coefficient becomes clearer with a little bit more maths (don't worry – software such as Excel JMB or Minitab can do it for you). The value r^2 shows the percentage of variation in Y explained by the effect of X. For example, if $r = 0.7$, the variable is causing 49 per cent of the variation in Y; if $r = 0.8$, the value increases to 64 per cent. In either of these circumstances, you seem to have found the important root cause of the problem as these values are especially high, particularly considering that a number of other Xs are also influencing the Y results. With a lower value, for example where $r = 0.2$ or 0.3, the impact is relatively small, accounting for 4 per cent and 9 per cent, respectively.

Figure 8-7 shows the line of best fit, which can help you see the likely values for data that you don't currently have. Drawing a line through highly correlated data such as that in the first two examples in Figure 8-6 is easy – you can do it with a ruler and pencil. You can calculate the line precisely using the regression equation, $y = b_0 + b_1x$, where b_0 = the intercept (where the line crosses the vertical axis, $x = 0$) and b_1 = the slope (the change in y per unit increase in x).

You'll see this equation presented in a number of ways, but whichever letters you use, the slope will look the same!

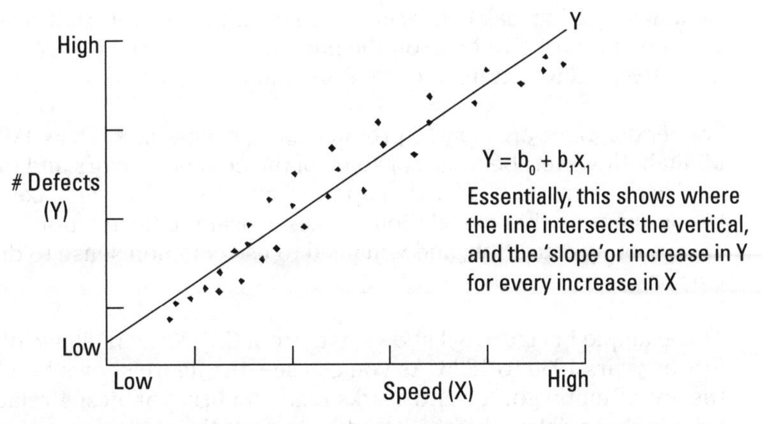

Figure 8-7: Working out the line of best fit. Here you can use the data to help you predict things, but remember the potential for a threshold point that changes the picture

When just one X is involved, this calculation is known as *linear regression. Multiple regression* extends the technique to cover several Xs, as does *design of experiments*, but these more involved statistical techniques are outside the scope of this book (take a look at *Six Sigma For Dummies* by Craig Gygi, Neil DeCarlo, and Bruce Williams and *Six Sigma Workbook For Dummies* by Craig Gygi, Bruce Williams, and Terry Gustafson, both published by Wiley).

Linear regression enables you to make predictions for the value of y with different values of x, though remember that the straight line might not continue forever. A future threshold may exist where things change dramatically, as we show in Figure 8-8.

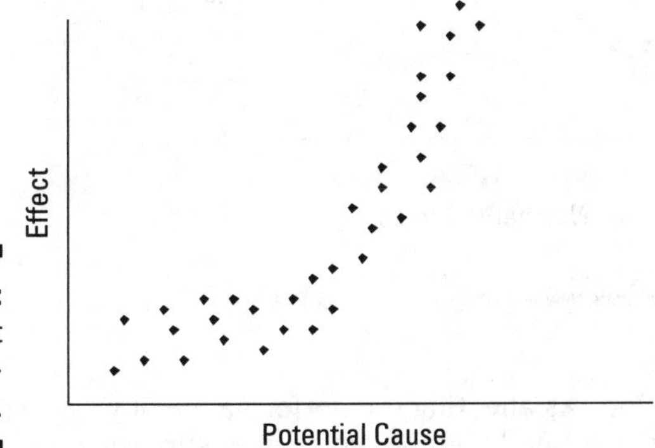

Figure 8-8: Looking out for thresholds.

Abandon rate in a call centre is a good example of a future threshold: callers might be prepared to hang on the line for a reasonable time, but at a certain point they become irate and slam the phone down.

Scatter diagrams are easy to produce using programs such as JMP, Excel or Minitab. However, be aware of some of the common errors and pitfalls associated with them, such as mixing up the X and Y variables and axes or making the assumption that correlation implies causation. Correlation does not always imply causation, and you need to use common sense to draw your conclusions.

The example in Figure 8-9 shows data from the German village of Oldenberg, for the years 1930 to 1936. As you can see, the figure shows that Walt Disney's Dumbo got it right: storks really do bring babies! A relationship does exist in these data – but the X and Y axes are the wrong way round. The village expanded in this period, people built new houses, and the increase in the number of tall chimneys proved to be an attraction for nesting storks. More usefully, we could plot the number of houses on the X axis and the number of storks on the Y axis.

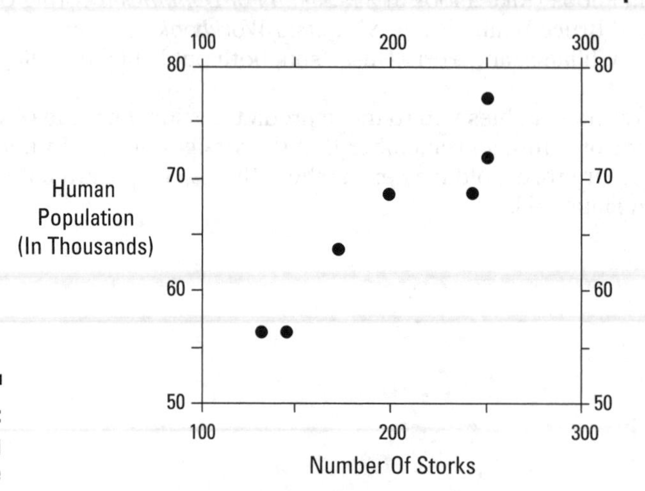

Figure 8-9: Bringing home the baby.

Understanding the various Xs affecting the performance of your process is crucial. Once you've identified your Xs, you can ensure the right measures are in place and can work towards creating stable and predictable performance.

Assessing your effectiveness

In analysing your performance, you may also want to put in place some additional measures, such as *overall process effectiveness* (OPE) in transactional processes and *overall equipment effectiveness* (OEE) in manufacturing. We use OPE and OEE to measure and understand the performance and effectiveness of equipment or processes. Three elements are important:

- The availability rate measures downtime losses from equipment failures and adjustments as a percentage of scheduled time.
- The performance rate measures operating speed losses – running at speeds lower than design speed and stoppages lasting seconds.
- The quality rate expresses losses due to scrap and rework as a percentage of total parts run.

These elements are multiplied together, where OEE = Availability × Performance × Quality. So, with Availability at 90 per cent, Performance at 95 per cent, and Quality at 99 per cent, the OEE = 0.90 × 0.95 × 0.99 = 84.6 per cent.

In service organisations, or for transactional processes, OPE tends to be used. Here, you take the following three elements, again multiplying them together to determine the OPE:

A = availability of equipment

P = productivity

Q = quality rate

Take a look at *Lean For Dummies* by Natalie J. Sayer and Bruce Williams (Wiley) for more detailed information about the OEE and OPE.

Part IV
Improving the Processes

'OK, dwarfs — there's going to be some changes around here.'

In this part . . .

A variety of tools and techniques come together to help you reduce waste and the time it takes to do things. In this part, we look at how to improve the process flow so that you can achieve results in less time and with less effort. We cover a number of concepts including 'pull not push', the power of prevention, and the importance of people issues in ensuring successful change.

We also give you the lowdown on Value-Added and Non-Value-Added, and look at the importance of identifying and tackling bottlenecks in your processes.

Chapter 9

Identifying Value-Added Steps and Waste

In This Chapter

- Adding value to your organisation
- Seeing how most problems result from just a handful of issues
- Working waste out of your organisation

'We need to add value.' How often do you hear someone in your organisation say something similar to this? Perhaps you use the phrase yourself. Unfortunately, many organisations don't have an agreed definition of 'value-added' or indeed 'non-value-added' and this leads to confusion and missed opportunity. People place different interpretations on what this commonly used Lean Six Sigma expression means, and the use of the terms to help remove unnecessary steps and actions, and to simplify processes, can be lost.

This chapter focuses on waste, generally, describing the 'seven wastes' identified by Toyota's Taiichi Ohno that need to be removed to reduce costs and processing time.

Interpreting Value-Added

Lean Six Sigma focuses on providing value for the customer (see Chapter 2), so knowing what value actually means in your organisation is crucial. Chapter 4 covers the CTQs, those critical to quality customer requirements that your organisation needs to meet; in examining how your processes try to meet those CTQs, you need to assess whether all the steps involved are really necessary. For determining if each step adds value to your process, a standard definition that everyone in your organisation can use and understand is a prerequisite.

Providing a common definition

For a step to be value-added, it must meet the following three criteria:

- The customer has to care about the step.
- The step must either physically change the product or service in some way, or be an essential prerequisite for another step.
- The step must be actioned 'right first time'.

The first criterion in this list is rather subjective. Put yourself in the shoes of the customer: if they knew you were doing this particular step, would they be prepared to pay for it? In providing value for your customer you need to give them the right thing, at the right time, and at the right price (see Chapters 2 and 4 on meeting CTQs).

You need to look at your process from your customers' perspective. You may be processing their orders in batches, for example, and waiting until you've completed the entire batch before despatching the products. The step putting an individual customer's order to one side while you finish processing others hardly adds value from their perspective.

Consider another example. You have to refer your customer's mortgage application to a senior underwriter to approve the loan. The customer won't be happy to pay for this step, especially if it involves sending their papers to another location – they expect you to be able to approve the loan. If the process involved the client's paperwork for the mortgage going back and forth between underwriters the situation would be even worse.

The second criterion – the step must change the product or service – means that activities such as checking, revising, expediting, and chasing are clearly non-value-added. Challenging your process steps with this criterion seeks to prevent unnecessary checking and the movement of items back and forth between different steps in the process.

Chapter 5 describes a process-stapling exercise for highlighting non-value-added steps. Some steps in your process may be completely unnecessary – so remove them. Ensure the removal won't cause an unexpected knock-on effect elsewhere in the process, though. If you carry out the process stapling thoroughly, you can see all the vital elements in your process and how they interrelate and you can make a simple improvement with no unforeseen adverse effects.

Very often, managers have 'bolted on' these unnecessary steps as a knee-jerk reaction to something going wrong in a process. It was almost certainly the wrong thing to do, but before too long, it became recognised as an important step in the process. As other errors occurred over time, so more 'bolt-on' steps were added, leading now to a prime opportunity for a value-added analysis to help make the process flow more easily. Our experience is that checking work to avoid errors can be a pretty hit-and-miss affair – we always recommend trying to build quality in through prevention and error proofing (see Chapter 10).

Making sure a step is done right first time is the third criterion in checking value-added. Re-work costs time, effort, and money and is definitely a non-value-added activity. Chapter 10 looks at addressing errors using prevention and error proofing.

Carrying out a value-added analysis

After you establish a common definition for value-added, you can review your processes and see if any non-value-added steps can be removed. This section describes how to go about a value-added analysis, but bear in mind that you're likely to want to keep some of the non-value-added steps you discover. For example, some regulatory requirements may be in place that the customer may not be interested in, but which you must adhere to.

You need to analyse your process, so looking at Chapter 5 may be helpful. A value-added analysis really is as straightforward as it sounds, though: just look at each step in your process and determine if it's necessary. Use the matrix in Table 9-1 to capture your data. Completing and analysing the detail might create some surprises; typically, very few steps add value.

Table 9-1 A Value-added Analysis

Process Step	*Unit or Activity Time*	*Value-Added Time*	*Non-Value-Added Time*
Vet application			
Enter on system			
Run credit check			
Issue offer			
Diary follow up			
Client confirms			
Issue cheque			
Total Time			
Percentage Time	100%		

As part of your analysis, assessing the unit or activity time for each of the process steps is sensible. Unit time is the time it takes to complete a process step (we cover unit time in more detail in Chapter 5). If you know how long a step takes to complete and the salary costs associated with the people working in the process, you can work out the approximate cost of that non-value-added activity, which may well encourage you to improve the step or eliminate it.

Understanding the unit time is relevant for all of the non-value-added steps, but perhaps especially so in terms of the re-work activity. Chapter 5 looks at mapping your processes. Very often process maps are produced assuming the work is carried out right first time. Unfortunately, this ideal situation isn't always the case, as you can see in Figure 9-1 (the dotted lines represent rework).

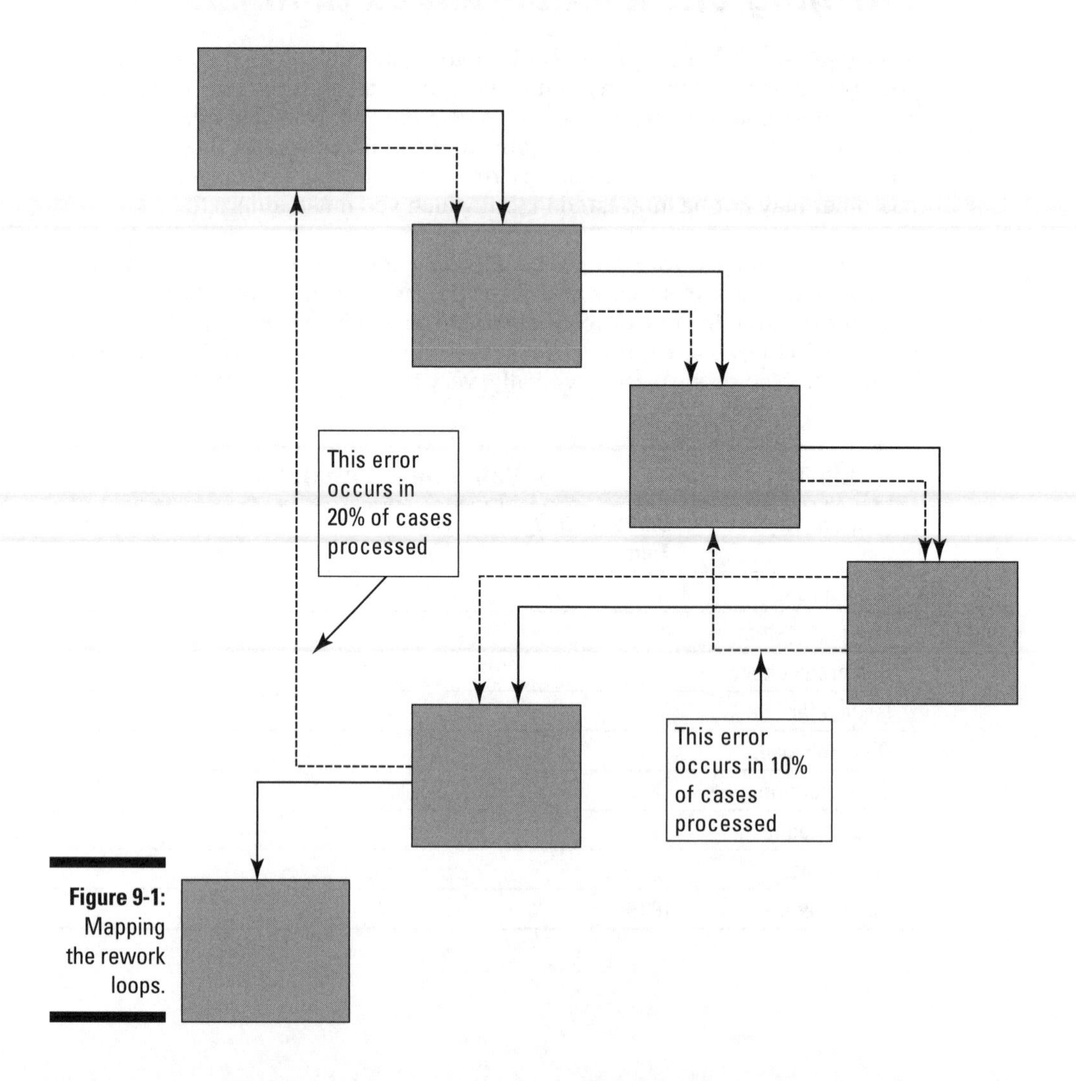

Figure 9-1: Mapping the rework loops.

Mapping, perhaps in a different colour, the re-work loops in your process and recording how often they're used can be very revealing. In possession of your cost information, you can then start to prioritise your efforts to prevent these expensive errors (see Chapter 10 for more on preventing errors occurring in the first place).

Once identified, many non-value-added tasks can probably be eliminated, but some will remain necessary for regulatory, or health and safety, or environmental reasons. Termed 'essential non-value-addeds', these activities need to be carried out as quickly and efficiently as possible. Ensure your process allows this to happen.

Assessing opportunity

Typically, only 10 to 15 per cent of the steps in a process add value, and more often than not, these steps represent as little as 1 per cent of the total process time. These numbers may sound impossible, but just think about what happens in your own processes.

In relation to these figures, the scope for improvement is huge, especially in reducing cycle time – the time it takes to process a customer order. In Chapter 1, we refer to Taiichi Ohno's quote:

> *All we are doing is looking at a timeline from the moment the customer gives us an order to the point when we collect the cash. And we are reducing that timeline by removing the non-value-added wastes.*

Reducing the timeline between a customer's order and receiving payment is your mission and the value-added analysis described in this chapter starts you on that journey.

Looking at the Seven Wastes

Muda is Japanese for waste. In any process, some steps add value and some don't. Some of these non-value-added steps have to stay, however, perhaps because of limitations in available technology or resources. Others can be eliminated immediately, perhaps through a DMAIC (Define, Measure, Analyse, Improve, and Control) project (see Chapter 2).

In waste terms, non-value-added steps are described as either 'type one' or 'type two' Muda. These broad types of waste can be broken down into seven categories:

- Overproduction
- Waiting

- Transportation
- Processing
- Inventory
- Motion
- Correction

Sometimes, these categories are introduced as 'Tim Wood':

- **T**ransportation
- **I**nventory
- **M**otion
- **W**aste
- **O**verproduction
- **O**ver-processing
- **D**efects

In the following sections we look at each of these wastes in turn, using these seven categories.

Owning up to overproduction

Overproduction is producing too many items, or producing items earlier than the next process or customer needs. This type of waste contributes to the other six wastes.

Working in a service organisation, we discovered a classic example of *process sub-optimisation*: improvement or inappropriate targets in one part of the process cause problems elsewhere in the process. The manager of Department A was determined to show how good both he and his team were compared with the other departments involved in the downstream process. He set a production target that required overtime work from his team, but which showed extremely high levels of productivity that earned praise from senior management. Unfortunately, this increase in output created problems in the immediate downstream process step, leading eventually to the work being stored as a two-week backlog. Even more unfortunately, the manager of Department B received the blame and was pressured by senior management and those working on the further downstream process steps. Overproduction had struck again!

A classic example of overproduction involves printed material. When you see how the unit price for leaflets or brochures, for example, dramatically reduces as the volume increases, over-ordering is really tempting. Ordering the higher volume and paying so much less per unit makes sense. Or does it? Do you have large amounts of printed material that is unlikely to be used, or is taking up valuable storage space, or is out-of-date? How much is literally thrown away?

Playing the waiting game

Waiting essentially means people are unable to get on and process their work. This delay may be due to equipment failure, for example, or because people are waiting for the items they need in their part of the process.

Waiting can be incurred by late delivery from external or internal suppliers, or perhaps the incomplete delivery of an order. One of our favourite examples of unnecessary waiting involves photocopiers. Large organisations need top-of-the-range photocopiers but often senior management decide, for reasons of cost, that basic models will suffice. Unfortunately, cheaper models aren't designed to deal with the volume demands of large organisations, and keep breaking down, leaving staff to either wait for them to be repaired or wander round trying to find a photocopier that does work.

Troubling over transportation

Transportation waste involves moving materials and output unnecessarily. Sending partly completed batches of work through the internal mail system because processing teams are located inappropriately is an example. Movement of goods to and from the same place is another. Transportation processes involving non-value-added steps are even more wasteful.

Transportation waste can also include moving surplus material (see the 'Owning up to overproduction' section, earlier in this chapter). The need to move things around in order to find space for other things, for example, is often the result of overproduction. (Chapter 11 covers pull production systems, whereby items are only requisitioned when they're actually needed.)

Picking on processing

Processing waste covers performing unnecessary processing steps, involving, for example, irrelevant information, or the completion of too many fields on a form. Putting too many additional information leaflets into a letter being sent to a customer – piggy-backing on the real purpose of the correspondence – is another example.

Processing of unnecessary information is the real crux of processing waste. Consider situations in which customers filling in order forms, and people processing them, have to provide or input more information than is really needed. Eventually, the processing team identify the 'key fields' and, provided they complete those, the application can be progressed. So what was the other information for? Sometimes it can be justified as potentially important marketing information, but so often it isn't needed or even used.

Investigating inventory

Inventory waste links to overproduction resulting in too much work in progress or too many brochures, leaflets, or stationery in stock.

Working in a bank, we reviewed the process for issuing cheque statements to account holders. Large machines are used to sort the statements and stuff the envelopes, but this particular work area didn't have sufficient capacity to deal with the volume of statements. They needed an additional machine, but had no room for one. Or did they?

In reality, poorly utilised space in their storage room (in part caused by keeping over-ordered and out of date literature) had created an overspill of inventory onto the work area floor space. This overspill was preventing the acquisition and location of the additional and much needed machine. We used the Five Ss – sort, straighten, scrub, systemise, and standardise (see Chapter 10 for more on these) – as a framework to help keep things neat and tidy, with a place for everything and everything in its place. Using the Five Ss, we created space to facilitate the introduction of the extra machine and an accompanying increase in processing output.

Watch out for the overproduction of items simply to meet productivity targets. A demand for costly space to store them is likely to result!

Moving on motion

Time for some ergonomics. Motion waste covers a range of movements, including that of people, perhaps due to the inappropriate siting of process teams or equipment, or the need to find misplaced documents. This type of waste also includes the need to access too many screens, double-handling, or seeking unnecessary approvals.

Motion waste certainly includes unnecessary movement caused by a poorly designed workspace: positioning of computer screens or the height of a desk or work bench, for example.

Some years ago, researchers compared the relative positions of the controls on a lathe with the size of an average male worker. They found that the lathe operator had to stoop and move from side to side to operate the controls. An 'ideal' person to fit the lathe would measure 4 feet 6 inches tall, 2 feet across the shoulders, and have an arm span of 8 feet!

This example epitomises the shortcoming in design when no account has been taken of the user. People come in all shapes and sizes, and ergonomics takes this variability into account in the design process. Ergonomics is about ensuring a good fit between people, the things they do, the objects they use, and the environments in which they work, travel, and play. A range of best practice guidelines is available, covering areas such as lifting and the ideal design of workstations.

One goal of ergonomics is to design jobs to fit people. Job design in ergonomics recognises everyone is different. Variability in height, weight, length of arms, size of hands, and so on, needs to be taken into account, and study of the human body (anthropometrics) gives data on how these vary across the population.

Applying these principles involves following a logical process:

1. **Analyse the job.** What is required to do the job properly and safely?
2. **Identify any stressful elements of the job, focusing on issues related to physical movement.** Is machine access too tight for the largest worker? Do short workers have to crane their necks to read displays? Do workers have to reach above their shoulders or below their knees?
3. **Determine the relevant body dimensions linked to the problems identified.** Height, weight, arm length, or hand size can be issues, for example.
4. **Decide how much variability needs to be accommodated in the design.** You can use data from various anthropometric studies (studies of the human body) to help you determine the appropriate specifications in your design. Many of these are available on the Internet or from relevant government departments. You might be able to create the design based on the actual measurements of your existing staff, for example, using average height and/or weight information.
5. **Involving the operators and users, redesign the workstation as appropriate.** Build in adjustment capability to accommodate size or arm length differences between staff.

Benefits of applying this five-step process are improved efficiency, quality, and job satisfaction. Costs of failure include error rates and physical fatigue, or staff absence through injury.

Coping with correction

Correction is the seventh waste and it deals with rework caused by not meeting customer requirements (the CTQs – critical to quality customer requirements), providing incomplete replies, or simply making errors.

Figure 9-1, in the 'Carrying out a value-added analysis' section earlier in this chapter, is a process map showing levels of rework. You can use unit time information to put a cost on rework. American quality guru Phil Crosby refers to PONC – the *price of non-conformance*. This simple measure puts a price on how much it costs to do things wrong, to not meet customer requirements. He estimates that in a service organisation, PONC is somewhere between 25–40 per cent of annual expenditure!

Errors are a costly waste – Chapter 10 focuses on how to prevent them.

Focusing on the Vital Few

Witnessing the scale of waste in your own organisation makes you appreciate the need to tackle things in bite-sized chunks. Quantifying the scale of the waste problem, and breaking things down into manageable pieces, involves measurement. Chapter 6 covers measuring your processes using check sheets and Pareto diagrams. Pareto's 80:20 rule won't always be exact, but a vital few Xs are likely to be causing most of the problem(s). You need to find these vital few.

Chapter 10

Discovering the Opportunity for Prevention

In This Chapter

- Applying good housekeeping
- Using prevention rather than cure
- Undertaking proactive maintenance
- Levelling your processes

The concept of prevention has been in existence for a long time – a very long time. Even before our grandmothers told us that prevention is better than cure, and probably even before Laozu had highlighted its importance back in 600 BC:

Before it moves, hold it,

Before it goes wrong, mould it,

Drain off water in winter before it freezes,

Before weeds grow, sow them to the breezes.

You can deal with what has not happened,

Can foresee

Harmful events and not allow them to be.

Prevention is a good way to tackle waste and delays in your processes (which we cover in Chapters 9 and 11). If you have less waste and fewer delays, you reduce the need for rework and avoid some other non-value-added activities, too. For example, by providing more information upfront to your customers, you may reduce the volume of customer enquiries.

Keeping Things Neat and Tidy

A simple 'housekeeping checklist' can help you to reduce waste and wasted effort in the workplace. In Lean Six Sigma, you use a system called *the Five Ss* to create this checklist.

In using the Five Ss, you need to run a *red tag exercise*, a tracking process that identifies things in the process that you don't need, whether inventory, stationery, or something else.

You may also find that visual management techniques, such as signs, symbols and colour coding help you in your housekeeping.

Introducing the Five Ss

The Five Ss link to the concept of just in time (see Chapter 11 for more on this approach) which aims to provide the tools and materials you need to do the job *only* when you need them. Implementing the Five Ss usually leads to a safer and more pleasant working environment that encourages both self-management and team working. Here are the Five Ss:

- **Sort** encourages you to look at the tools, materials, equipment, and information you need to do your job, and separate them into those used 'frequently', 'occasionally', and 'never'. You can sort based on your experience, but 'tagging' the items in some way can be helpful (see the 'Carrying out a red tag exercise' section later in this chapter).
- **Straighten** literally means straightening things up and putting everything you use frequently easily to hand. Straightening may include, for example, toolkits, files, or email folders, or may involve moving a printer to a more convenient location. Things that you don't use frequently need to be put somewhere else or thrown away! You need to decide how many items need storing, how they should be stored, and where. Naturally, these stored items should be appropriately labelled to facilitate their easy access in the future.
- **Scrub** concerns keeping the things you use, and the environment you work in, clean and tidy, and appropriately maintained. Make your workplace shine: get rid of rubbish and dirt, and don't leave scrap lying around. Make sure your tools are current, safe, and clean, and that all the information and documents you use are up-to-date and well presented. Check that equipment and machinery are routinely serviced and maintained.

- **Systemise** means strengthening your approach. Design a simple way of working so that your tools and information stay sorted, straightened, and scrubbed. Essentially, systemise involves regularly re-doing the first three Ss! Doing so helps identify the reasons why the workplace becomes messy and cluttered, and prompts preventive thinking to find ways of stopping the problems recurring.
- **Standardise** the whole approach and keep doing it. Stick to this system every day, train everyone in the application of the Five Ss, regularly review things, and tell others about your effective method of working, so it becomes a way of life.

Carrying out a red tag exercise

A *red tag exercise* is a tracking process to help highlight unneeded items. If you use the Five Ss, which we describe in the previous section, red-tagging can become a useful element of 'Sorting'. You could, for example, tag the various items on your desk on a particular date and then see when you use them next, updating the tag with the time and date. If you haven't used them in, say, one month, then move to 'Straighten', so they can be appropriately relocated. Once all the obvious things have been thrown away, or 'relocated', you'll be left with only those things that you regularly use and need to hand.

You can use the red tagging at home, too – for example, to keep your wardrobe from bursting.

You may need to form a team to work on red-tagging your wider working areas, appointing a champion and team members. You identify the areas to tackle, for example inventory, equipment, stationery, and supplies, and agree and communicate the criteria and timeframe of the exercise.

The team red-tags items, evaluates the results, and agrees and takes the necessary actions. Working in this way involves organising the needed items regularly so they're easy to locate and use. The team may use the red tag exercise at the same time as introducing or enhancing visual management (see the next section) to make checking that items are where they should be easier. Visual management helps ensure that items in use can be returned to the right place, and that missing items are easily identified; see in the 'shadow board' for tools in Figure 10-1.

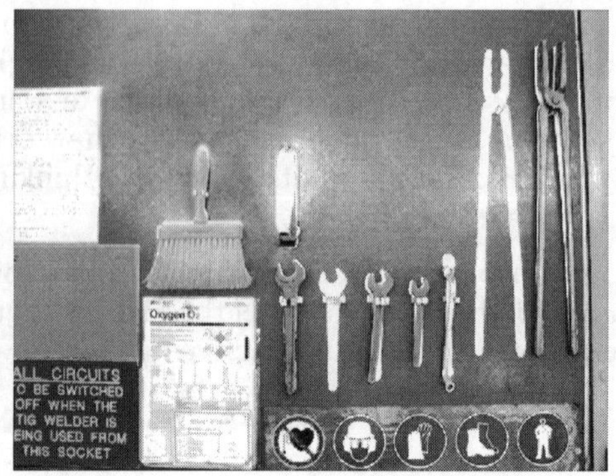

Figure 10-1: A shadow board helps you see at a glance if any tools are missing.

Using visual management

Examples of visual management are everywhere, and most people see them every day, as road signs, traffic signals, and written notices. Visual management aims to:

- Help people know where they are.
- Highlight important messages and rules about health and safety.
- Help people know where to find things.
- Highlight when things aren't where they should be.
- Identify when things go wrong – and what to do.
- Communicate information about performance, to both staff and visitors.
- Help people prevent errors and accidents by using standard colours.
- Help to highlight waste.

In the workplace, visual management helps keep things well organised, and ensures things can be found easily. This approach also offers a way to communicate performance results, both local and corporate, for example, ensuring that people understand how things are going.

Visual management takes many forms, from using standard colours for pipes, cables, and wiring, to showing clearly marked out walkways or parking spaces, and designated floor space for equipment and machines.

Visible management isn't enough by itself, though. Management attitudes, behaviours that lead to appropriate and timely actions, and use of the other preventive tools and techniques that we describe in this chapter need to support the approach.

Looking at Prevention Tools and Techniques

You can prevent or at least reduce the impact of problems by using a whole range of tools and techniques. Sometimes, they'll be all you need to achieve Lean Six Sigma performance.

Introducing Jidoka

Jidoka is a Japanese word describing the prevention of defects and it works on the principle that once a defect or error occurs, no further defects or errors are produced until the cause of the problem is remedied. In 1902, Sakichi Toyoda, the founder of the Toyota group, invented an automated loom that stopped each time a thread broke. This immediate halt prevented the thread spewing out and so saved time that previously was wasted in sorting out the ensuing mess. A printer stopping when its ink runs out is a modern example of Jidoka.

Without Toyoda's concept, automation has the potential to allow a large number of defects to be created very quickly, especially if processing is in batches. Jidoka is often referred to as *auto**no**mation* – a means of preventing defective items from passing to the next process.

Autonomation allows machines to operate autonomously, by shutting down if something goes wrong. *Automation with human intelligence* is another term for this concept. We highlight the 'no' in autonomation to remind you that *no* defects are allowed to pass to a follow-on process.

Jidoka also embraces the concept of 'Stop at every abnormality' which means a manual process stops whenever an abnormal condition occurs. Sometimes in manufacturing, every employee is empowered to 'stop the line', perhaps following the identification of a special cause on a control chart (see Chapter 7 for more on these).

Forcing everything to stop and immediately focus on a problem may seem painful at first, but is an effective way to quickly get at the root cause of issues. In batch processing, discovering problems immediately is crucial.

Reducing risk with Failure Mode Effects Analysis (FMEA)

FMEA is a prevention tool that helps you identify and prioritise potential opportunities for taking preventive action. Identifying the things that might go wrong – the *failure modes* – is the first step.

By looking at what might go wrong, you can assess the impact of what happens when it does go wrong, how often it is likely to occur, and how likely you are to detect the failure before its effect is realised. For each of these probabilities, you assign a value, usually on a scale of 1 to 10, to reflect the risk. Table 10-1 provides a typical rating scale for a service organisation.

Table 10-1 Weighing up the risk

Rating	*Severity of Effect*	*Likelihood of Occurrence*	*Current Detectability*
1	None	Remote	Immediately detected
2	Very minor effect	Very low	Found easily
3	Minor	Low	Usually found
4	Low to moderate	Low to moderate	Probably found
5	Moderate	Moderate	May be found
6	Moderate to high	Moderate to high	Less than 50% chance of detection
7	High	High	Unlikely to be detected
8	Very high	Very high	Very unlikely to be detected
9	Hazardous	Extremely high	Extremely unlikely to be detected
10	Disastrous	Almost certain	Almost impossible detect

To help you prioritise your actions, you calculate a *risk priority number* (RPN). This value is the result of multiplying your ratings for the severity of the risk (from Table 10-1), the frequency of occurrence, and the likelihood of detection. You then find ways to reduce the RPN.

In determining ratings for the various failure modes in your processes, working with members of the relevant process team and looking at each step in the process is sensible. To ensure you identify each step, use a deployment flowchart or value stream map (see Chapter 5 for how these work). Your ratings against the descriptions in Table 10-1 are based on your experience rather than absolute fact, so when you complete the exercise, step back and make sure the numbers seem sensible. Next, prioritise the failure modes that need to be addressed and determine actions to reduce the RPN scores.

Consider a pharmacist as an example. Going back a few decades, many things could go wrong – for example, a child could open a bottle of their parents' tablets, swallow them, and become ill. Looking at the ratings in Table 10-1, the severity rating is therefore high. The detection rating is also high and the occurrence rating is probably somewhere in the middle. This failure mode may happen every day, but the pharmacist is unaware until the damage has occurred.

The RPN in this example is high. To address the failure mode, pharmacists now use child-proof containers and print warnings on them about keeping tablets out of the reach of children, create tablets that don't look or taste like sweets, and reduce the strength of individual tablets.

Examine your own processes to see if FMEA creates any opportunities for improvement. Consider each step in your process and identify its failure modes. In coming up with your RPN, remember that these numbers are subjective; use common sense in determining the action needed.

In creating your list of possible failure modes, call upon your own experience but also use techniques such as *negative brainstorming* (also called anti-solution brainstorming). This technique turns brainstorming on its head and instead of asking, for example, 'What possible failure modes are evident in the ABC process?' takes a different tack and tries, 'How can we ensure the ABC process goes wrong?'

You may be surprised by the ideas that your team members put forward. Some of the suggestions will be silly, but that doesn't matter – indeed, it helps make the exercise fun.

When you have your list of failure modes, ask the question, 'How many of these things really happen?' Changing your negative statements to positive ones can produce a solution to your problem.

Negative brainstorming not only helps develop your list of potential failure modes, it can also begin the process of identifying ways to reduce the risks and improve the process.

Error proofing your processes

Error proofing – sometimes referred to as *Poka-yoke*, Japanese slang for 'avoiding inadvertent errors' – is key to working out the actions you need to take to improve your process.

Poka-yoke approaches either prevent mistakes from being made or make the mistakes obvious at a glance. Poka-yoke approaches are:

- Inexpensive
- Very effective
- Based on simplicity and ingenuity

Poka yoke doesn't rely on operators catching mistakes, but it does help to ensure quick feedback 100 per cent of the time, leading to process improvements and reductions in waste.

Consider the 1-10-100 rule, which states that as a product or service moves through the production system, the cost to your organisation of correcting an error multiplies by 10. Looking at the processing of a customer order, for example:

- Order entered correctly: £1
- Error detected in billing: £10
- Error detected by customer: £100

The 1-10-100 rule fails to pick up the additional costs associated with dissatisfied customers sharing their experience with others. Error proofing, all in all, is really worth doing.

Examples of prevention and error proofing are observable in your everyday life. Your car may flash a warning light if you don't use the seat belt. Some high-tech cars even have breath-testing devices that prevent ignition if you exceed the legal alcohol limit. And at home, you probably have smoke detectors and electricity trip switches.

Three types of error proofing approaches exist: contact, fixed value, and motion step.

- **Contact error proofing** involves products having a physical shape that inhibits mistakes – see Figure 10-2.

  The physical design makes installing parts in any but the correct position impossible. Electronic equipment design, and that of its various attachments and extensions, for example, ensures the right cables can only go into the right sockets. This situation is achieved through a combination of part sizes and shapes, as well as colour codes. Another example is a fixed diameter hole through which all products must fall. Any oversized product is unable to pass through, and the potential defect associated with it is thus prevented.

- **Fixed value error proofing** identifies when a part is missing or not used and essentially ensures appropriate quantities. A simple example is the French fry scoop used in a takeaway, which is designed to ensure a consistent number of fries precisely fits the package served to the customer. A further example is 'egg trays' used for the supply of parts – spotting that something's missing is easy as one compartment is empty.

- **Motion step error proofing** automatically ensures that the process operator has taken the correct path or number of steps, possibly by breaking a photocell light sensor, or stepping on a pressure sensitive pad during the assembly cycle. A very different example is how spell-checkers provide automatic warnings when words are incorrectly spelt throughout the completion of a word processing document – the operator needs to click on the highlighted word to decide whether to change it.

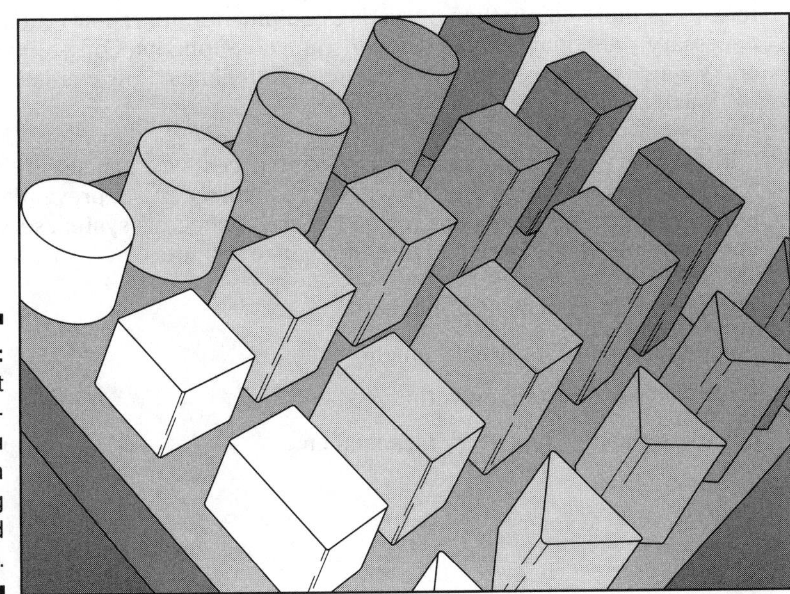

Figure 10-2: Contact error proofing: You couldn't fit a square peg in a round hole.

Profiting from Preventive Maintenance

Preventive maintenance means being proactive to prevent equipment failure and system problems. Contrast this approach to diagnostic or corrective maintenance, which is performed to correct an already-existing problem. If you have and look after a car, you may understand the concept of preventive maintenance: you don't change your oil in response to a problem situation – you do it before things go wrong, so your engine lasts longer and you avoid car troubles down the road.

Preventive maintenance is a schedule of planned maintenance actions aimed at the prevention of breakdowns and failures. Preventing the failure of equipment before it actually occurs is the primary goal. Preventive maintenance is designed to preserve and enhance equipment reliability by replacing worn components before they fail and activities include equipment checks, partial or complete overhauls at specified periods, oil changes, lubrication, and so on. In addition, workers can record equipment deterioration so they know to replace or repair worn parts before they cause system failure. Recent technological advances in tools for inspection and diagnosis have enabled even more accurate and effective equipment maintenance.

An ideal preventive maintenance programme prevents all equipment failure before it occurs. For example, in an airport, preventative maintenance may be in place in critical service areas such as escalators, lighting, and aircraft bridges.

As with all prevention activity, some people see preventive maintenance as unduly costly. This logic dictates that regular scheduled downtime and maintenance is more costly than operating equipment until repair is absolutely necessary – and may well be true for some components. Long-term benefits and savings associated with preventive maintenance, however, also need to be considered.

Without preventive maintenance, for example, costs for lost service time from unscheduled equipment breakdown will be incurred. Also, preventive maintenance results in savings due to an increase of effective system service life. Long-term benefits of preventive maintenance include:

- Improved system reliability.
- Decreased cost of replacement.
- Decreased system downtime.
- Better spares inventory management.

You can't always prevent things from going wrong or equipment from failing. But when they do, your ability to recover from problems quickly is key.

Avoiding Peaks and Troughs

This section focuses on dealing with work activity to avoid too many peaks and troughs in the volumes and types of work being processed. Levelling the work isn't easy, but it is the foundation of Toyota's celebrated production system. The Japanese refer to the concept as *Heijunka*, extending the concept to incorporate the need for 'standard work' – the processing of work in a consistent way.

Introducing Heijunka

Heijunka is the underlying concept of the Toyota Production System (TPS), shown in Chapter 1 in Figure 1-1. The TPS consists of two columns – Jidoka (for more about this concept, see the 'Introducing Jidoka' section earlier in this chapter) and just in time (see Chapter 11) – supported by Heijunka.

Heijunka involves smoothing processing and production using levelling and sequencing. For a process to run smoothly and consistently with many different kinds of output, it has to average, not just in volume, but also in kinds. So, you need to process the different types of customer order, for example, based on the date they're received rather than dealing with the more straightforward cases first and allowing the more difficult ones to build up and be delayed.

Heijunka involves the following elements:

- **Levelling** involves smoothing the volume of production in order to reduce variation. Amongst other things, this technique seeks to prevent 'end-of-period' peaks, where production is initially slow, but then quickens in the last days of the sales or accounting period, for example.
- **Sequencing** involves mixing the kinds of work processed. So, for example, when setting up new loans, the type of loan being processed is mixed to better match customer demand, and help ensure applications are actioned in date order. Managing this approach may be easier in manufacturing, where a producer may be able to place a small buffer of finished goods near shipping, to respond to the ups and downs in weekly orders. Keeping a small stock of finished goods at the very end of the value stream, this producer can level demand to its plant, and to its suppliers, making for more efficient utilisation of assets along the entire value stream while meeting customer requirements.

- **Stability and standardisation** seeks to reduce variation in the way the work is carried out, highlighting the importance of following a standard process and procedure. This technique links well to the concept of process management and the control plan, where the process owner continuously seeks to find and consistently deploy best practice.

Concepts such as Heijunka can't be implemented overnight – for example, Toyota has taken many years to achieve the successful application of levelling and spreading the load but is now a driving force in the growing awareness of lean thinking principles in the western world.

Spreading the load

Keeping things balanced and level means your process flows are smoother and your processing times faster. But be warned, this situation isn't easy to achieve, either at work or on your way there!

Consider variable speed limits on motorways, which aim to maintain a steady, continuous flow of traffic, enabling us to all keep moving and avoid stops and starts during busy periods. Unfortunately, some drivers always speed up between the speed cameras, only to brake hard when they get to the next one, an approach that creates braking and delay back down the road.

In the workplace, you need to try to avoid peaks and troughs in activity, if you can. The month or quarter end cycles in many organisations highlight the difficulties of peaks. Actioning financial reconciliations, for example, on a daily or weekly basis may be possible, thus avoiding the monthly or quarterly peak of activity. You need to determine whether an opportunity exists in your organisation to change frequencies.

In Chapter 9 we talk about waste, or *Muda*. This expression is often used together with two other words, Mura and Muri. *Mura* describes unevenness in an operation, for example people hurrying then having to wait, as in the motorway scenario. *Muri* means overburdening equipment in some way.

Consider Mura and Muri in the context of maintaining a smooth and level flow at a transport depot. You have several three-tonne trucks, but you need to transport six tonnes of material to your customer. You have four options:

- All six tonnes on one truck = Muri and a probable broken axle.
- Four tonnes on one and two on another truck = Muri, Mura, and Muda.

- Two tonnes on three trucks = Muda.
- Three tonnes on two trucks = Muri-, Mura-, and Muda-free!

So, this last option is the optimum way of delivering the material to your customer. It uses an evenly distributed approach, no waste occurs, and trucks aren't over-burdened.

Carrying out work in a standard way

Sometimes, the first step in preventing problems and rework is agreeing on a standard process. Formulating a standard process gives you real gains easily and leads to stability and predictability in the process. Actually, you can't really begin to improve a process until you standardise it. Following standardisation, you have a genuine chance to stabilise the process and prompt further improvements. In Chapter 5, we look at techniques such as process stapling and process mapping to help you develop both the best approach and a standard approach to the way work is done.

Standardising the 'one best way' of how the work gets done is key, but in a culture of continuous improvement you may find better ways to do the work that become the new 'one best way', until further improvement occurs. Of course, if defects occur, your first question needs to be, 'has the standard process been followed?' If it has, then the process needs to be improved.

In this evolving culture of continuous improvement, fuelled and supported by Lean Six Sigma, you need to keep improving your process, encouraging ideas from the people working within it. As you grow increasingly confident in applying the Lean Six Sigma principles, so you'll recognise there's no end to the process of improving processes!

Chapter 11

Identifying and Tackling Bottlenecks

In This Chapter

- Finding the weak points in your process
- Picking pull instead of push production
- Rethinking the layout of your organisation

In this chapter we focus on those points in the process flow where demand exceeds capacity – bottlenecks. In Lean Six Sigma, some people call bottlenecks 'constraints'. Whichever term you use, the effect is the same: the bottleneck, or constraint, sets the pace for your process and determines the speed and volume of your output. Put simply, either you manage the bottlenecks or they manage you.

Applying the Theory of Constraints

You need a framework to help you manage your constraints. This section looks at how to identify the bottlenecks in your process, prioritise them for action, and reduce or eliminate their effect using Eli Goldratt's five-step approach.

Identifying the weakest link

Think of your organisation as a chain like the one in Figure 11-1 – a series of functions or divisions that are dependent on each another, even if the people within the organisation don't recognise and accept that fact. For example, you don't ship parts until they're packaged, and you don't package parts until they're manufactured, and so on. Answering the question 'How strong is the chain?' is easier than you might think: the chain is as strong as its weakest link. Find your bottleneck and you find the weakest link in your chain.

Figure 11-1: Working on the chain gang.

Conventional wisdom supports the idea that improving any link in the chain improves the chain overall, and 'global' improvement is the sum of the local improvements. But time for some different thinking: this local improvement approach leads to *process sub-optimisation*, where apparent improvements in one part of the process actually make things worse in another part. You need to make your improvements with an understanding of the end-to-end process, or the chain. And you need to take only those local actions that strengthen the chain, by focusing potentially scarce resources on the constraint.

Improving the process flow

Eli Goldratt suggested a *theory of constraints* involving a five-step approach to help improve flow:

1. Identify the constraint.
2. Exploit the constraint.
3. Subordinate the other steps to the constraint.
4. Elevate the constraint.
5. Go back to Step 1 and repeat the process.

A *constraint* is a bottleneck. A constraint occurs wherever and whenever capacity cannot meet demand. You can *identify* constraints where you have a build-up of people (a queue), material (inventory), units to be processed, or work in progress (a backlog).

When you find the bottleneck or constraint, you can then find ways to improve the processing capability at the bottleneck point in the process flow. You need to *exploit* the constraint, that is, maximise its potential.

For example, if your constraint is a machine, try to keep the machine running during the working day. Don't close it down for servicing: you can service the machine after hours. Any time lost at the constraint has a big effect on the whole process, which takes you to Step 3 of the theory.

To *subordinate* the other steps to the constraint, you use the constraint to dictate the pace at which the upstream activities send their output to the constraint, and which tells the downstream activities how much they can expect to receive from the constraint.

As an example, consider the deployment flowchart in Figure 11-2, featuring Ann, Brian and Clare.

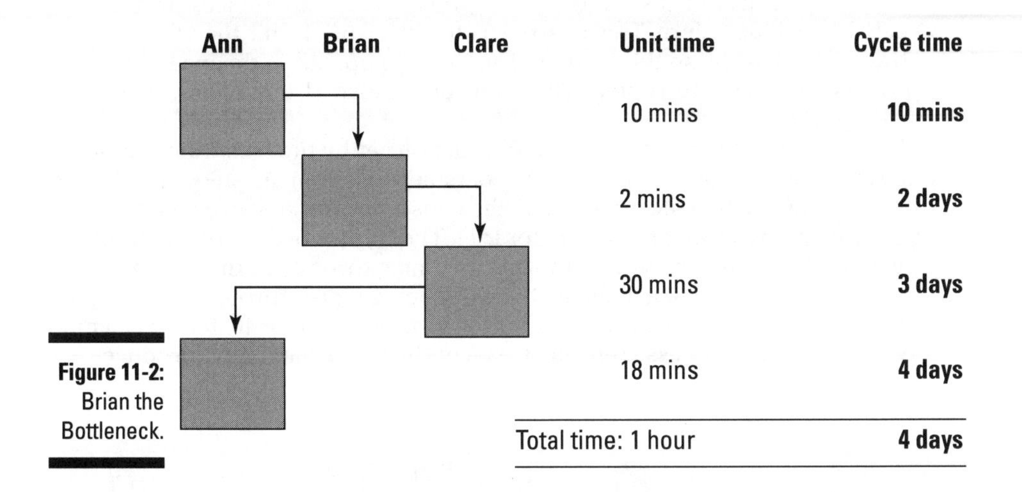

Figure 11-2: Brian the Bottleneck.

Brian is the main bottleneck. Anna producing more than Brian can deal with is pointless, as doing so will create an increasing pile of work in progress. Brian is setting the pace for the process and ideally, is *pulling* the work through accordingly, rather than letting Ann *push* it through at her pace. This situation may mean that Ann can tackle some additional tasks, possibly even helping Brian. Either way, you can improve the constraint.

To elevate the constraint means to improve it. You can introduce improvements that remove this particular bottleneck, possibly through a DMAIC (Define, Measure, Analyse, Improve, and Control) project – Chapter 2 takes you through the steps. Of course, once you initiate changes, a new constraint will appear somewhere else in the flow, so you start this improvement cycle again.

Step 5 takes you back to the beginning to *repeat* the process – a route to driving continuous improvement. In this example, you would then need to address the bottleneck with Clare.

Building a buffer

The constraint sets the pace for the process – it tells the upstream process steps how much to produce and the downstream process steps how much to expect. However, imagine if one of the upstream steps wasn't able to produce things on time – a machine could break down, for example. Placing a small buffer in front of the constraint to ensure sufficient work is always available is a good idea, just in case an upstream process step experiences problems, such as machine failure. This upstream step can work faster than the constraint if necessary, so things should soon catch up, but in the meantime the process flow is uninterrupted. This concept is called 'drum, buffer, rope'.

The imaginary 'drum' is the beat of production set by the constraint, rather like the drum beating the pace for the oarsmen on a Roman galley. The buffer provides the contingency that keeps the constraint working even if one of the upstream steps slows or fails temporarily. The 'rope' cordons off, or controls, the flow of work by preventing too much coming through to the constraint – this image also helps you imagine the work being pulled through at the right pace. In Figure 11-3, the drum equates to the production of 40 items each day, even though the process steps upstream of the constraint could produce more.

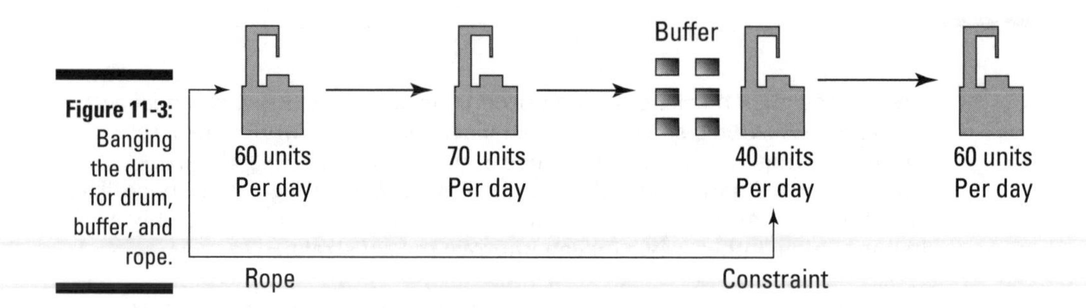

Figure 11-3: Banging the drum for drum, buffer, and rope.

Managing the Production Cycle

Whether you work in a manufacturing organisation or a service industry, you need to understand and manage the production process.

Using pull rather than push production

Pull production is a system in which each process takes what it needs from the preceding process exactly when it needs it, and in the exact amount necessary. The customer thus pulls the supply and helps avoid being swamped by items that aren't needed at a particular time. In our example in Figure 11-2 in the preceding section, Brian pulls the work through at his pace when he wants it, not when Anna can send it.

Pull production reduces the need for potentially costly storage space. For example, in an environment where pull production isn't in use, overproduction in one process, perhaps to meet local efficiency targets, may result in problems downstream, increasing work in progress and creating bottlenecks. Symptoms of overproduction include the following:

- **Too many:** Making more items than needed.
- **Too soon:** Making them earlier than needed.
- **Too fast:** Making them faster than needed.

Pull production links naturally to the concept of *just in time*. Just in time provides the customer with what they need, when they need it, and in the quantity demanded. This concept applies to both internal and external customers, but it demands a very closely managed relationship with suppliers, something that can take many years to achieve.

Ideally, the downstream activities signal their needs to the upstream activities through some form of request, for example a kanban card (*kanban* is Japanese for a card) or an electronic andon board. An *andon* can be a light that flashes when more stock is required, for example, or you could use a card signalling that a goods-in tray is empty or at a certain level and needs topping up.

Whatever signal is agreed, nothing is produced upstream until the request is made and a signal is flagged. If the activity is being processed within a 'cell' (see the 'Using cell manufacturing techniques' section later in this chapter), seeing and managing the pull operation is more straightforward as everyone is working closely together.

A simple example of the kanban signal in practice is the stationery cupboard. A re-order card is placed in an appropriate position within the stock and when the card is revealed as someone takes a new memo pad from the remaining pile, for example, a re-order is made to ensure the stock of memo pads doesn't run out. The kanban system is also evident in your chequebook, as a re-order form towards the end of it.

Moving to single piece flow

Single piece flow means each person in the organisation performs an operation and makes a quick quality check before moving their output to the person in the next process. If a defect is detected, Jidoka is enacted (this concept is covered in Chapter 10); that is, the process is stopped, and immediate action taken to correct the situation and take countermeasures to prevent reoccurrence.

Ideally, single piece continuous flow processing is carried out within 'cells', with the relevant people and machinery sited closely together. See the 'Using cell manufacturing techniques' section later in this chapter.

Recognising the problem with batches

Single piece flow is a real change of thinking that moves you away from processing in batches. Traditionally, large batches of individual cases or items are processed at each step of the process and are passed along the process only after an entire batch has been completed. Delays are increased when the batches travel around the organisation, both in terms of the transport time, and the time they sit waiting in the internal mail system – at any given time, most of the cases in a batch are sitting idle, waiting to be processed. In manufacturing, this idleness is seen as costly excess inventory.

In batch processing, errors can be neither picked up nor addressed quickly. If errors occur, they tend to occur in large volume, which further delays identifying the root cause. In single piece flow, the error is picked up immediately. With the trail still warm, you can get to the root cause analysis faster and prevent a common error recurring throughout the process.

Looking at Your Layout

In many organisations, especially in offices, the various people involved in a process often aren't sitting together, and can even be located on different floors or in different buildings. This type of layout inevitably results in delays as people or work travels around the organisation.

Identifying wasted movement

People and materials lose vast amounts of time travelling between different locations (Chapter 9 covers waste and how to eliminate it). Use the spaghetti diagram in Chapter 5 (Figure 5-1) to help you reduce wasted travel time.

Using cell manufacturing techniques

Cellular manufacturing organises the entire process for similar products into a group of team members, with all the necessary equipment. This group and its equipment is a *cell*. A cell shouldn't feel like a prison; rather, it should feel liberating for the team members who have real control over what they produce.

Cells are arranged to easily facilitate all operations, often adopting a horse-shoe shape, as shown in Figure 11-4. Outputs or parts are easily passed from operation to operation, often by hand, eliminating set-ups and unnecessary costs, and reducing delays between operations.

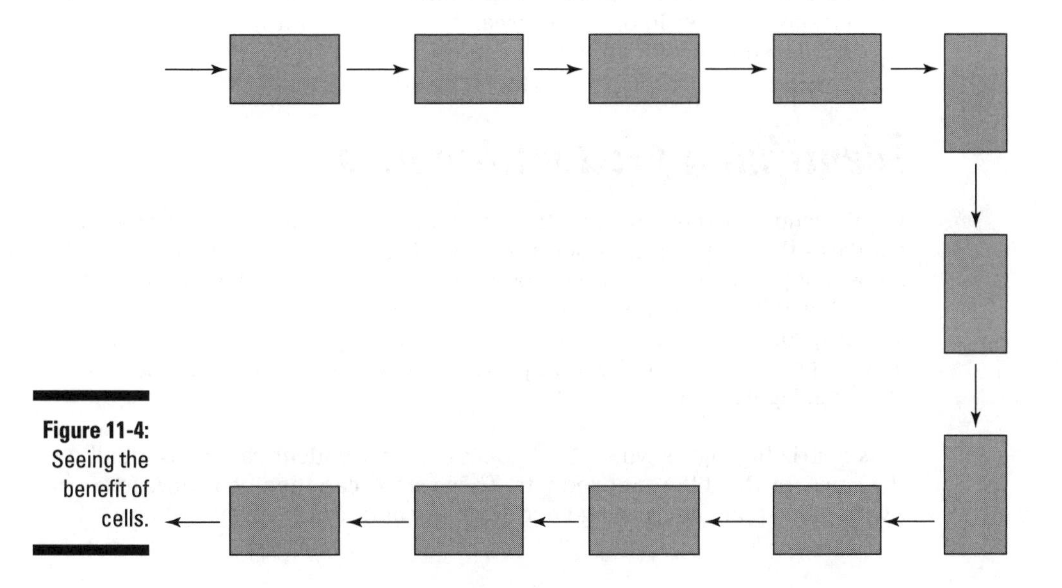

Figure 11-4: Seeing the benefit of cells.

Working in cells offers the following benefits:

- The facilitation of single piece flow and a reduction in the use of batches.
- Faster cycle times and less work in progress, which in turn results in a need for less floor space.
- Reduction of waste and minimisation of material handling costs resulting from less movement of people and materials.
- More efficient and effective utilisation of space.
- A heightened sense of employee participation.
- More efficient and effective use of people in the team, empowering them to take responsibility and control (see Chapter 12). Daily team meetings are easier to arrange, helping to create a culture of continuous improvement, which results in a positive team attitude and an enrichment of job satisfaction.
- Elimination of bottlenecks.
- Facilitation of visual management – with everyone in the process working closely together in the same area, the team's performance results can be displayed easily for all to see.

Identifying product families

Within cellular arrangements, identifying and processing common *product families* – those products or services involving identical or similar processing steps that previously might have been seen as different activities, each processed by different teams – makes sense. To identify the appropriate product families, you need to create a matrix detailing the process/value stream steps across the page and the different products or services down the page, as shown in Figure 11-5.

This matrix highlights where the process steps are identical or essentially the same for the different products. These steps can then be processed by the same team, increasing your flexibility and processing capability.

	Vet application	Enter on system	Run credit check	Issue offer	Diary follow up	Client confirms	Issue cheque
Bronze plus	X	X	X	X	X	X	X
Silver edge	X	X	X	X	X	X	X
Gold	X	X	X				X
Platinum	X	X					X
Platinum plus		X					X

Figure 11-5: Keeping it in the family.

Chapter 12

Understanding the People Issues

In This Chapter

- Gaining buy-in from others
- Tackling resistance to change
- Clocking the culture of your organisation

Six Sigma and Lean originated from industrial manufacturing backgrounds, with early emphasis on tools and techniques. Now, however, most managers accept that recognising and handling the people issues is the biggest challenge in implementing Lean Six Sigma successfully.

Lean Six Sigma aims to make change happen in order to improve things. Human beings, like most creatures, are cautious and sceptical about change – it spells danger. We humans have an inbuilt resistance to change, especially if somebody tells us it is going to be 'good for us'. Most people fear losing something they have as a result of change.

Dealing with personal fear and loss is another big challenge in implementing Lean Six Sigma, but few enthusiasts in statistical theory cover this in their extensive training.

Understanding people is key to implementing a Lean Six Sigma project. Almost always, if Six Sigma and Lean projects fail, people issues of one form or another are the cause. In this chapter we offer guidance and tips for managing the human aspects of change in Lean Six Sigma.

Working 'Right' Right from the Start

Unfortunately, we don't know an easy formula for solving the challenge of managing people in a Lean Six Sigma project. However, in over 80 implementations of Lean Six Sigma, we've found a small number of common factors that consistently stand out as critical for success. Perhaps not surprisingly, leadership commitment is one of these critical factors. Clinching buy-in at the beginning is the real challenge.

As part of an extensive study of Lean Six Sigma projects, Andy Liddle, a well-known guru in Lean Six Sigma in Europe, identifies three high-level essentials for successful deployment:

- **Right work:** Ensure your projects are focused on the right issues and linked to your business objectives. Find out what your organisation considers important and then use your organisation's business strategy to drive your Lean Six Sigma projects.
- **Work right:** Run your projects effectively, using good-quality people, applying sound project management techniques, and ensuring rigorous governance through committed sponsorship, project reviews, and tollgates (refer to Chapter 2 for more on tollgates).
- **Create the right environment:** Lean Six Sigma flourishes as the natural way of working in certain settings. Creating the right environment is about leadership, recognition, encouraging people to do the right things effectively, ensuring the environment supports the team's work, and putting in place controls to avoid people sliding back into their old ways.

Gaining Acceptance

Overcoming the resistance movement in your organisation may be one of the biggest challenges you face when you introduce Lean Six Sigma. You may find many people are reluctant to accept this new way of working (especially with the suspiciously confusing name of Lean Six Sigma). You can't ignore or get away from these people: resistance exists in all organisations, no matter how big or small, and at all levels.

Making change happen successfully in an organisation is difficult. John P. Kotter, a respected expert in organisational leadership, researched 100 organisations that failed in their first business transformation attempt. In his *Harvard Business Review* article 'Why transformation efforts fail', Kotter identifies eight common failure factors:

- Not establishing a sufficient sense of urgency.
- Not creating a powerful enough leadership coalition.
- Not creating a vision.
- Under-communicating.
- Not removing obstacles to the vision.
- Not systematically planning for and creating short-term wins.
- Declaring victory too soon.
- Not anchoring changes in the culture.

In the next section, we show how you can use these negative thoughts to your advantage in your organisation.

Managing change

Turning the negative statements that we describe in the previous section into positives provides the basis for a model for managing change as shown in Figure 12-1. You can use this model during the life of each Lean Six Sigma project and also across the entire Lean Six Sigma programme for your organisation. Work from left to right in the model starting with establishing the need for change.

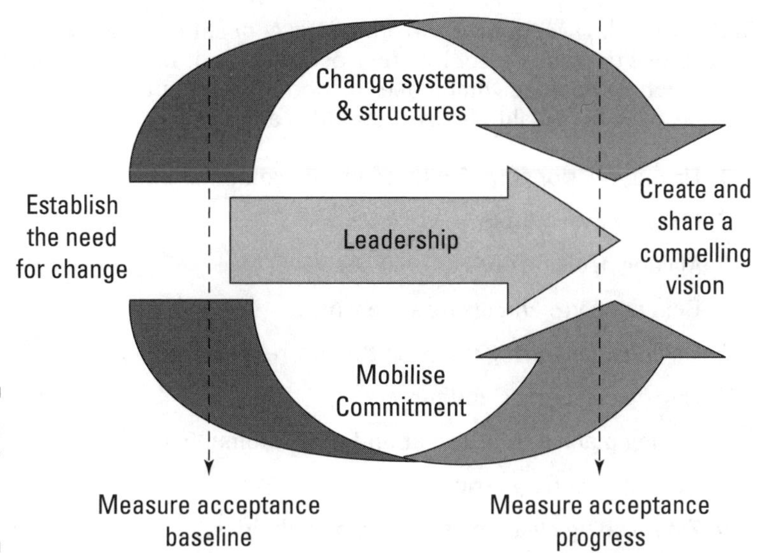

Figure 12-1: A model for managing change.

George Eckes, a well-known writer on this subject, uses a simple but eloquent expression to express gaining acceptance for change and overcoming resistance, whether for a whole Lean Six Sigma programme or for the changes resulting from part of a Lean Six Sigma project.

$Q \times A = E$

Q is the technical quality of the solution: The 'hard' tools of Lean and Six Sigma will have proven that the solution works when tested.

A is the acceptance of the change by people: Having a high 'score' for A is as important as having a good-quality solution.

E is the effectiveness of the change in practice: The solution actually works in practice.

Some hardened practitioners believe that the A factor is more important than the Q factor and is the real key to success in Lean Six Sigma. To understand how people perceive things and to win support, you need to score well on both factors.

If you're in the early stages of deploying a Lean Six Sigma programme, the A factor is likely to start with winning support from senior managers.

Keep Q × A in mind as a simple shorthand for a highly complicated issue, dealing with the human mind.

Overcoming resistance

In the preceding section we introduce the concept of a good A 'score' – or acceptance of your project by other people in your organisation. We could write a whole book on what makes a good A score, but Estelle Clarke, one of Europe's leading Quality figures, provides a useful guide in this list:

1. Having relationships built on mutual commitment.
2. Leading by example.
3. Making decisions based on facts.
4. Being open to change how we work.
5. Having an inspiring vision of the future.
6. Having clear goals and targets.
7. Having plans that are clear and well-communicated.
8. Having a winning strategy.
9. Establishing clear roles and responsibilities.
10. Dealing effectively with those resisting change.
11. Having no need to fire-fight.
12. Attracting and retaining world-class people.
13. Following world-class processes.
14. Having clear and open communication.
15. Learning from each other.
16. Working well together across all functions.

17. Encouraging teamwork.
18. Finding more productive ways of working.
19. Having consistency of purpose.
20. Having a high sense of urgency.
21. Making decisions quickly.
22. Continuously seeing to achieve competitive advantage.
23. Continuously building customer confidence.
24. Using measures to compare our performance with best practices.
25. Being honest and sincere.
26. Understanding our market, customers and competitors.
27. Facing up to problems quickly.
28. Rewarding the right behaviours.
29. Encouraging creativity and innovation.
30. Modifying systems and structures to support business assurance.
31. Achieving budgeted objectives.
32. Doing what we say we will do.
33. Expecting our performance standards to increase continuously.
34. Encouraging expressions of different points of view.
35. Only reporting relevant information.
36. Wanting to learn from our mistakes so we don't repeat them.

You can map the list on to the change model in Figure 12-1 and use it as the basis for an organisational assessment involving a simple scoring mechanism. Simply adding the words 'How good are we at . . .' before each phrase can help you develop a questionnaire. For example, with number 17 on the list you can ask 'How good are we at encouraging teamwork?'.

In practice you can adapt the list for a specific project and add an assessment score, typically using 5 levels: 1= very poor, 2 = weak, 3 = fair, 4 = very good, 5 = excellent. This technique can be useful to measure the 'acceptance baseline' in Figure 12-1. Don't feel you have to use a questionnaire for every project though; simply interviewing and listening to people who have an interest in the project (the stakeholders in the project) using these questions can be a really useful approach.

Creating a Vision

Clear, 20:20 vision may well be something that organisations try to develop in the next ten years. Talking about 'visions' in a book about Lean Six Sigma may seem rather fanciful – but visions help you paint a picture that appeals to people's hearts and minds and can help you answer the question 'Why change?'

Customers, business leaders, and employees all view the future from different perspectives. Imagining a time machine is an ideal way to develop a vision: you can speed ahead and discover for real what it will be like when the change has been completed. What is different? What is there more of and less of? Being in the future you can find out how the change has affected people's attitudes and behaviours. What does it feel like now? How does it look from the view of the customer, the leader, and the employee?

A time machine is outside the scope of even the most extensive Lean Six Sigma Master Black Belt toolkit, so you can use a simple technique called *backwards visioning*. This technique helps you create a picture of the future expressed in behavioural terms – that is, what the culture will be like in the future. The improvement team (the Lean Six Sigma team) imagines that their change has been completed successfully and then considers what they'd expect to see, both internally and externally, in terms of the following:

- Behaviours
- Measures
- Rewards
- Recognition

By determining the team's perceptions of these issues, you can begin to understand the actions that you may need to take as part of your progress towards the desired state: the future after the change has been made. These actions include the activities and behaviours that you need to reduce and remove and those that you need to introduce and increase.

You may want a more supportive culture, where people help each other more, work in a team more, and operate less on their own private agendas. For example, as a manager you see a piece of litter in the corridor. Do you walk on because you didn't drop it and cleaning this area isn't your job? Or do you pick it up and set an example? Creating a vision is about leadership, taking responsibility, and working in the best interests of the business.

Writing down a backwards visioning statement provides a helpful framework for developing influencing strategies. For example, a good vision for the future for an airport operator working on reducing queues and increasing security is:

> *Our goal is to transform the security experience of the travelling passenger by: (a) exceeding expectations by eliminating queues, and (b) creating a highly professional environment overseen by security staff who are rigorous, professional, helpful, and proactive.*

A clear vision provides clarity about the outcomes of the change effort and helps you to identify at least some of the elements that the change aims to transform. A vision secures commitment and support from anyone involved in delivering this service by helping people understand what you want to change – and why.

Understanding Organisational Culture

Defining the concept of 'culture' in organisations is difficult – yet most people have an idea of what the term means in their own organisation. They know their organisation's 'unwritten rules' and can describe 'the way things get done around here' a lot more vividly than a written rulebook or set of documented policies can.

At the core of most organisations is a set of values and beliefs that pervade everything and dictate more strongly than any management fad what people *think* should be done and *how* it should be done. These enduring beliefs create attitudes and behaviours that may undermine your Lean Six Sigma project if people consider your project to threaten them.

Alongside the formal declared change (the plan), another process is happening – often hidden in the shadows but still having a powerful impact. The best Lean Six Sigma practitioners recognise the hidden cultural, unwritten rules and manage change in the cultural process as keenly as they manage the work process being improved.

Many change initiatives fail because their proponents don't have enough awareness of the cultural factors involved. Many mergers and acquisitions fail to achieve the promised gains for this reason.

Culture is complex, powerful, and based on events of the past. In any organisation, rituals, stories, myths, heroes, and villains play important roles. Gerry Johnson at Cranfield University in the UK, developed the idea of a *cultural web*, shown in Figure 12-2. Essentially, the cultural web is 'the way we do things around here'.

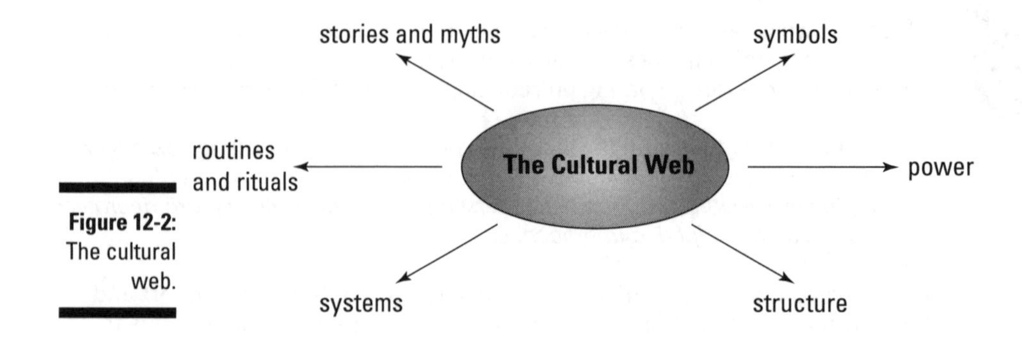

Figure 12-2: The cultural web.

Symbols in organisations may be much more significant than you think. Status symbols, language, and jargon may be based on associations with power and status.

Busting Assumptions

Assumption busting is a useful tool for challenging why things are done in the way they are. Keep asking 'why are things done this way?' to get beyond the initial, often superficial, responses.

A project to speed up registrations for land registry in the UK provides a good example. Registrations had always been sent to the legal department for review if the property value exceeded a particular amount. It transpired no logic was involved in this decision as the value of the property made no difference to the complexity of the title – but it did add three months to the completion time!

Assumption busting is a quick and easy technique that works well in workshop style sessions with groups or simply as a tool to work through on your own. Question the obvious things with a fresh pair of eyes. Ask why have we always done it that way?

Here are some common assumptions:

- It's impossible to do that.
- The rules won't allow that approach.
- We'll never get it through IT in time (okay, maybe that one is fair!).
- Department A, B, C (take your pick) will never agree.
- These all need to go for authorisation before being processed.
- Every project needs to have all 164 project documents produced before being given the go-ahead (even small projects?).

- We don't have the money, equipment, room, or personnel.
- Central office would never agree to it.
- You can't teach an old dog new tricks.
- It's too radical a change.
- That's beyond our responsibility.
- The employees will reject it outright.

You need to make everyone aware that these assumptions exist and be brave enough to challenge the status quo.

Seeing How People Cope with Change

Several models illustrate the stages that people go through when coping with change in their lives. Many people know the Kubler-Ross model, involving the following stages: shock, denial, awareness, acceptance, experimentation, search, and integration.

Figure 12-3 illustrates how people typically react to change over time.

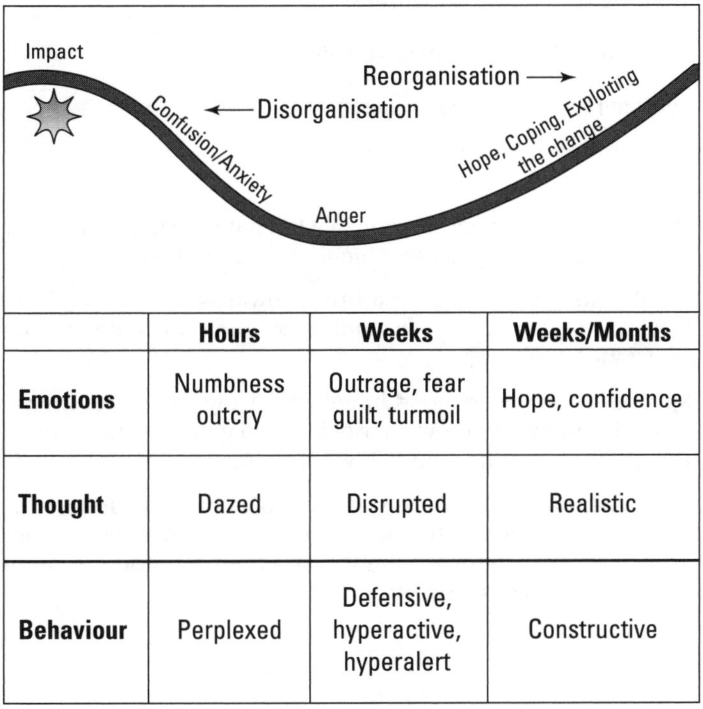

Impact
Reorganisation →
← Disorganisation
Confusion/Anxiety
Anger
Hope, Coping, Exploiting the change

	Hours	Weeks	Weeks/Months
Emotions	Numbness outcry	Outrage, fear guilt, turmoil	Hope, confidence
Thought	Dazed	Disrupted	Realistic
Behaviour	Perplexed	Defensive, hyperactive, hyperalert	Constructive

Figure 12-3: Change reaction.

Lean Six Sigma projects are about changing things for the better. You're trying to improve processes – so change is inevitable. Blindly hoping that doing the same things in the same way will magically improve your product or service is head-in-the-sand (HITS) thinking – otherwise known as being an ostrich. Unsurprisingly, although HITS thinking is a surprisingly common management practice, it's not taught in management schools or on Lean Six Sigma training courses!

Comparing energy and attitude

Not everyone reacts to the prospect of change in the same way, so looking at different responses is helpful. An 'energy and attitude scale', as shown in Figure 12-4, is a useful way of assessing people's attitudes.

SPECTATOR Positive Attitude Low Energy	**WINNER** Positive Attitude High Energy
DEADBEAT Negative Attitude Low Energy	**TERRORIST** Negative Attitude High Energy

Figure 12-4: Energy and attitude model.

- **Winners** are open and responsive to change. They want to do their best and have the energy to see things through to the end.
- **Deadbeats** have a negative attitude towards change and low energy levels. For them, change is a nuisance, and they undertake tasks with reluctance.
- **Spectators** have very good intentions. They have a positive attitude towards change, but low energy levels. Typically, they say the right things, but find it hard to follow through.
- **Terrorists** have high levels of energy but their attitude towards change is negative. Typically, terrorists have their own agenda. They can be very outspoken in their opinions, and their attitude can be summed up as 'that will never work'.

Fortunately, most organisations have a lot more winners than terrorists. By developing a strategy to tackle the 'soft stuff' (handling people) and respecting people's views and feelings, tackling change can be successful in even the most challenging organisations.

Using a forcefield diagram

You can take this model further by looking at the strength of support for, or resistance to, the outcome of a specific project. You can do this for different interest (stakeholder) groups using a forcefield diagram, as shown in Figure 12-5.

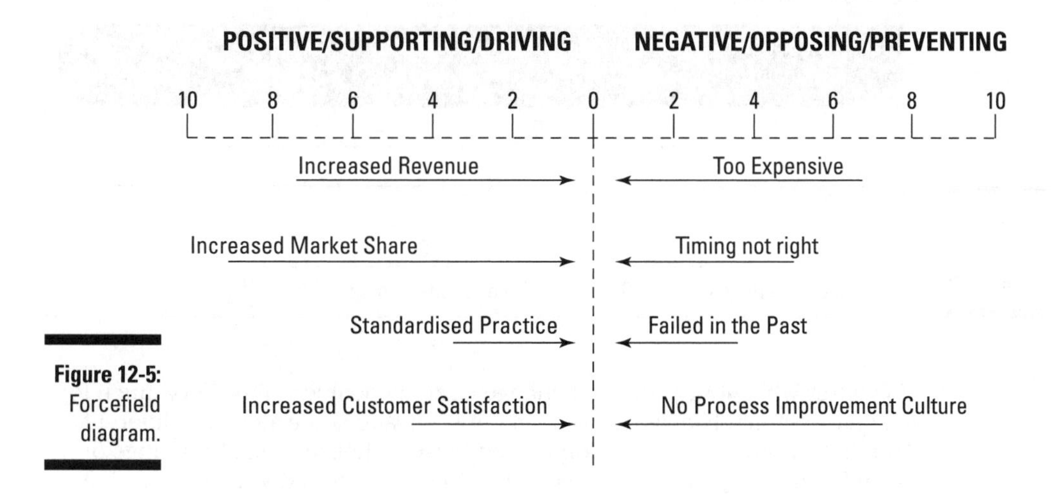

Figure 12-5: Forcefield diagram.

A forcefield diagram is a useful graphical representation of the positive and negative 'forces' influencing a particular project. You could also use a forcefield diagram when starting to implement an entire Lean Six Sigma programme in your organisation, as this diagram indicates. The length of each line represents the strength of the force. In Figure 12-5 the programme is perceived to bring increased revenue which is seen as a positive force, but this has to balanced by the negative force of being too expensive.

These forces are perceptions rather than facts, but will certainly influence stakeholders' opinions so understanding what these forces are, and how strong they are, can be very helpful in understanding why people feel the way they do.

Analysing your stakeholders

Stakeholder analysis is another useful technique for identifying the 'interest groups' (stakeholders) in your project and their levels of support.

Use the matrix in Figure 12-6 to show where the stakeholders are now on the positive/negative scale, and also where you'd like them to be in the future. This kind of stakeholder analysis needs to be regularly updated during the life of a project, and is best kept for the team's eyes only. You're dealing with sensitive stuff – how people think and whether they're for or against change.

Stakeholder matrix

Names	Strongly against	Moderately against	Neutral	Moderately supportive	Strongly supportive	Hot/Cold spots	Next steps
A	X		0				
B				X	0		
C					X 0		
D			X	0			

X = where they are   0 = where we need them to be

Figure 12-6: Stakeholder analysis.

No matter what your proposed change is, some people will be very much for it, some completely against, some in-between, and some almost indifferent. And that's pretty much life! Don't be surprised when you find this range of attitudes applying to your project, or to the solution that you and your team eventually develop. Finding out early in the project what the for/against situation looks like, both on the surface and beneath it, is a good idea.

A *key stakeholder* is anyone who controls critical resources, who can block the change initiative by direct or indirect means, who must approve certain aspects of the change strategy, who shapes the thinking of other critical parties, or who owns a key work process impacted by the change initiative. So, in your Lean Six Sigma team, ask:

- Who are the key stakeholders?
- Where do they currently stand on the issues associated with this change initiative?
- Are they supportive and to what degree?
- Are they against and to what degree?
- Are they broadly neutral?

Given their status or influence on your project, where do you need the stakeholders to be? Moving some stakeholders to a higher level of support may be both desirable and possible, so work out how you can do that. Consider what turns them on or off, and think about how you can present the project in a more appealing and effective way for them.

Part V
The Part of Tens

In this part . . .

Every *For Dummies* book includes a fun and informative Part of Tens. Here, we give you information on some of the excellent resources covering Lean Six Sigma that are available. We also provide a collection of lists that include best practices, common mistakes, and where to go for help.

Chapter 13

Ten Best Practices

In This Chapter

- Applying Lean Six Sigma in everyday operations
- Using common sense and being practical
- Keeping things simple

You can apply the Lean Six Sigma toolkit in organisations of all sizes and sectors. The following sections highlight details to take care of to ensure success.

Appoint a Deployment Manager

The deployment manager is appointed by the senior team to programme manage the Lean Six Sigma deployment across the organisation. Even if you have the best intentions, your Lean Six Sigma programme stands little chance of success unless you appoint a deployment manager to make this happen.

The Lean Six Sigma deployment manager's main tasks include:

- Designing the overall programme
- Planning the initial roll-out
- Engaging stakeholder support
- Setting the framework and structure
- Organising training and support
- Reporting on progress, targets, and measures
- Dealing with internal communications
- Sharing internal best practice

A deployment manager's role is often underestimated. They should be able to focus on getting the programme off to a good start by organising the selection of initial projects that will bring early tangible benefits, engage with the wider organisation, establish a suitable training programme, and achieve early results, thus increasing acceptance within the senior executive group.

As the programme develops, regularly reviewing how well the overall Lean Six Sigma programme is progressing, and comparing your approach and progress against best practice, makes sense. A small number of specialist organisations can provide this service, providing you with an independent audit report showing how your organisation matches up with others at the same stage.

Appreciate that Less Is More

Keeping things simple can be surprisingly challenging in a world where processes and systems can become overly complicated and not matched to changing customer expectations and requirements.

Businesses that haven't discovered Lean Six Sigma seem to have a built-in tendency to overcomplicate processes by inserting non-value-added steps as a result of one-off failures being treated as 'common cause' events.

In one organisation we're familiar with, a high-value sales quotation was issued incorrectly. After that one-off mistake, a senior finance manager insisted that every new quotation be sent for additional checking and authorisation before being despatched. As time progressed and more mistakes continued to happen, even further checking and inspecting steps were added into the process in the false belief that this would reduce the defect level. Customers then complained that they did not receive quotations in the time they expected. When we tackled this issue shortly after the organisation decided to use Lean Six Sigma, one customer stated: 'Your internal processes are like a black hole.' Applying Lean Six Sigma resulted in a simplified, faster process, a subsequent increase in customer satisfaction, and a reduction in customers switching to competitors. Improving the process wasn't difficult, but no one had ever looked at the process from either the view of the customer or across the whole organisation.

Adopting the strategy of keeping it simple is a good idea for any Lean Six Sigma project. Try using deployment flowcharts (explained in Chapter 5) to illustrate the unnecessary complexity inadvertently designed into processes across your organisation – you may well discover a vast number of checks and balances added into the overall process. Simply map the process across the entire organisation and count the number of different people involved in the process and number of crossover points. Then see how you can consolidate or reduce them.

Build in Prevention

Prevention really is better than cure. This adage applies in business just as in health – with the added benefit in business that a prevention-based system costs a lot less than a cure.

Imagine a situation where an organisation ships poor-quality products to customers, or doesn't clean its hotel rooms adequately, or sends out inaccurate or late invoices to clients. The cost of putting right these failures is much higher than preventing the failure in the first place. Handling customer complaints, carrying out expensive rework, and paying for additional warranty shipping are non-value-added costs that your competitor won't incur if they've fully adopted Lean Six Sigma. Losing customers adds the final insult to injury, as word spreads of your company's poor quality.

Many high-quality manufacturing plants use error detection and error prevention techniques such as Poka-yoke (Japanese slang for 'avoiding inadvertent errors' – see Chapter 10), but service and transaction-based organisations still see such techniques as a novelty. Error detection and prevention systems are inexpensive and highly effective at preventing errors occurring in everyday work.

For example, a handful of hotel chains now use Lean Six Sigma, and travellers who use several hotels can really tell the difference in quality. Think of the effect of finding a hair on the pillow just after you arrive for a stay at a premium brand hotel. By focusing on the process of cleaning the room, a Lean Six Sigma business almost completely eradicates the possibility of such a defect occurring, in the same way that the likelihood of an aeroplane engine failing is now miniscule.

Building quality into your processes and preventing failure before it can happen is the key to the success of companies like Toyota. And taking into account the cost of failure, actually means you incur less expense overall.

Challenge Your Processes

Understanding how the work gets done, and then improving processes, is at the core of Lean Six Sigma. We don't see the word 'process' as synonymous with heavy documentation, bureaucracy, or sluggishness – although we know that the word 'process' often conjures an image of a constraining rather than a liberating force.

The famous statistician, W. E. Deming, used to say that the key role of managers, is to 'change the processes':

> *Eighty-five percent of the reasons for failure to meet customer expectations are related to deficiencies in systems and process … rather than the employee. The role of management is to change the process rather than badgering individuals to do better.*

Take the lid off any organisation, look inside, and you'll find that the organisation is made up of a series of interconnecting processes. Using the Lean Six Sigma toolkit in a continuing cycle of assessing, improving, and maintaining this organisational system by challenging the state of these processes keeps the organisation fit and capable of consistently meeting customer requirements.

Go to the Gemba

The Japanese expression 'Go to the Gemba' means go to the place where the work gets done. This approach is used throughout leading organisations, such as Toyota, where senior managers are almost indistinguishable from shop floor workers as they continually support and encourage everyday Kaizen (continuous improvement) activities. (Chapter 2 explains Kaizen in more detail.)

To find out how the work really gets done in your organisation, you need to go to the place where it happens. Lean Six Sigma projects get to the root of problems by actually going to the workplace and involving the people who do the work. All too often, the process map in the company quality manual is fictional – it doesn't represent reality. A real process stapling exercise – walking through the process viewed from the thing being processed – is an eye opener. (For more on process stapling, check out Chapter 5.)

Recognise and be prepared to accept that significant differences may exist between the real life 'as is' process in the Gemba, compared to the 'should be' process, let alone the ideal 'to be' process.

Many organisations are now institutionalising time for senior managers to regularly 'Go to the Gemba'. After all, how can you win a round the world yacht race if you lose touch with what the crew are having to deal with everyday?

Manage Your Processes with Lean Six Sigma

Lean Six Sigma is traditionally used as a method for improving processes but you can also use the tools to help you manage your business processes on an everyday, ongoing basis. During the control phase of a Lean Six Sigma project, all the key ingredients of process management need to be put in place, including the measurement system needed to monitor the ongoing running of the process on a day-to-day basis.

In essence, the control phase leaves the process running sweetly, and with the following in place:

- A clear, customer-focused objective reflecting the CTQs (check out Critical To Quality in Chapter 1).
- An agreed process map (plug into process mapping in Chapter 3).
- An agreed data collection plan with an appropriate balance of X and Y measures (we delve into data collection plans in Chapter 8).
- An ongoing control plan (we cover control plans in Chapter 2).
- A standardised process with appropriate documentation in place (see more about standardisation in Chapter 10).
- A visual management system to be updated and used in practice on a day-to-day basis using statistical control charts when needed (see our vision of visual management systems in Chapter 10).

By running a series of DMAIC projects across the processes that are at the core of running your organisation, you put in place the basis for an effective process management system. (We get down with DMAIC in Chapter 2.)

Don't spend ages putting a detailed process management system in place before you run a DMAIC project. Although such a system may seem like a sensible and logical approach, in reality your business leaders may lose patience as months or years pass by and they see little or no change. Lean Six Sigma's whole rationale is *to make a difference*!

Pick the Right Tools for the Job

A newly trained Lean Six Sigma practitioner may want to use all their newly learned tools – but in practice the Pareto principle (in this case that 80 per cent of projects can be completed successfully by using only 20 per cent of the LSS tools) applies in the choice of appropriate tools for different projects. (Chapter 6 explains the Pareto principle in more detail.)

Here are a few tips for using the right tools:

- Consider that, used in the wrong way, even the best tools give bad results.
- Know what rigour is necessary and when.
- Remember that excellent influencing skills are as important as superb tools (see the Q × A = E equation in Chapter 12 for assessing the success of your project in relation to people's commitment to your ideas).
- Don't be tempted to disappear into your own analysis paralysis.
- Use the methodology and the tools together. Even the most rapid improvement event benefits from using the define, measure, analyse, improve, and control stages. We've seen a DMAIC exercise carried out successfully in a one-day rapid improvement workshop. (Chapter 2 has a full explanation of DMAIC.)
- Remember that 20 per cent of the tools will be used on 80 per cent of projects, so don't try to shoehorn every tool into a project.
- Keep the scope of your project simple and understandable.
- Don't be afraid to ask for help if you need to use a tool that you haven't used since training. Everyone needs support from time to time.

Tell the Whole Story

Most people like a good story. People learn a great deal through reading, listening to, and exchanging stories. Keeping a 'storyboard' or log of a Lean Six Sigma project is an excellent way to communicate the project to the wider organisation and to pass on what you discover.

Organisations implementing Lean Six Sigma often don't bother with storyboarding. Many people miss the importance of this technique as they strive to meet project objectives. Ignoring the use of storyboarding is shortsighted, however: If you don't capture the discoveries, challenges and 'Aha!' moments of a project, the rest of the organisation is none the wiser and potentially makes the same discoveries and overcomes the same challenges over and over again.

Lean Six Sigma storyboarding is a straightforward technique to record the knowledge gained from a project. Just like the taste of strawberries is preserved in a good jam, your storyboarded knowledge lasts for many years.

Intranet sites make a good storage area, offering easy access to such 'bottled knowledge' across the organisation. These sites have the added advantage that, unlike jam, they're not eaten but can be used many times over! You have no excuse for losing stories from your records.

The latest storyboarding techniques use a combination of slides, words, interviews, and videos to capture the essence of what happened in the life of a project. Of course, a full-scale TV style documentary isn't necessary. Simply writing up learning points and recording key events in the project story on a flipchart and taking digital photos of the various sheets can form a very useful record while the team is live and thoughts are easily captured. Lean Six Sigma projects will probably soon be featuring on Internet video site YouTube!

Hail the Role of the Champion

Projects are more successful if they relate to key business issues and everyone realises their importance.

A *champion* is a senior sponsor who provides support, direction, and financial and people resources to Lean Six Sigma, demonstrating the company's commitment to the approach and providing a direct link to company strategy.

To make Lean Six Sigma work in practice, you need to put in place two different champion roles:

- A Lean Six Sigma programme business champion
- A champion for every Lean Six Sigma project

Looking at the Lean Six Sigma programme business champion

Visible commitment from the Lean Six Sigma business champion demonstrates that senior management take the Lean Six Sigma approach seriously. In an ideal world, the most senior executive in the organisation has the role of overall programme champion, but in practice, a member of the senior executive management team is often a good choice.

Senior management needs to communicate its support to the whole organisation, showing that it treats the approach seriously and not as just another fad.

Jack Welch, perhaps the world's most famous advocate of Lean Six Sigma when he was chief executive of General Electric, is often cited as the ultimate role model business champion. When asked how to gain leadership commitment in organisations that don't have a 'Jack Welch at the top,' he replied, 'It's no good giving this role to Harry from Quality'. With all due respect to the excellent Harrys who work in quality departments, commitment has to be actively driven from leadership at the top of the company.

Perusing the role of the project champion

Every project deserves a champion who's prepared to devote the time and support needed to help the project team overcome any roadblocks on their journey.

The project champion is involved in selecting the project and the team members for it. As the project progresses, the project champion stays involved by:

- Providing strategic direction to the team.
- Developing and agreeing the project charter (refer to Chapter 2), ensuring the scope of the project is sensible.
- Staying informed on the project's progress and taking an active involvement in project reviews.
- Providing budget and resources to the project team and helping to ensure the business benefits are realised in practice.
- Being prepared to stop a project. We hope that this is an infrequent event, but some projects simply have to stop. This may happen at the end of the Measure phase if the facts and data demonstrate that the 'problem' doesn't exist (or more likely that a different problem has been identified and a new project needs to be started).
- Helping to get buy-in for the project across the organisation.
- Ensuring appropriate reward and recognition for the project team for their success.

Use Strategy to Drive Lean Six Sigma

Implementing a Lean Six Sigma programme is pointless if it isn't aligned with the direction being taken at a strategic level by the business. Lean Six Sigma is about making change happen and strategy is about deciding which direction the company is heading – so they need to work in tandem. Lean Six Sigma helps you deploy strategy within the operational business. Many businesses now use Lean Six Sigma techniques as an essential component of their wider business transformation programme. These organisations use the essential tools of the Define, Measure, Improve, and Control phases (which we describe in Chapter 2) simultaneously across multiple processes to create a transformed business with the right set of services and products being delivered through streamlined processes.

Chapter 14

Ten Pitfalls to Avoid

In This Chapter

- Avoiding the temptation to shoot from the lip
- Knowing when to stop analysing and start implementing
- Steering clear of project traps and doing the wrong things right

Of course, you want your Lean Six Sigma programme to be a big success. The approach has been around for a while now, so you can draw upon a wealth of experience; some good and some not so good. This chapter describes things that can go wrong so you can avoid the common pitfalls. We share our experience of observing many different organisations and building up a knowledge of what works and what doesn't. Unlike some doomsters who never seem to have a good word for anything, we certainly don't want to put you off Lean Six Sigma. So read on and see if these pitfalls are ones that are likely to affect you.

Jumping to Solutions

Many managers seem hard-wired to jump straight to a solution when presented with a problem. In action movies, everything works out in the end; the hero makes the right decisions in a split second, and the goodies live on for another day (or film). Unfortunately, business life isn't quite the same: knee-jerk solutions can be costly and can fail to address the root cause of the problem.

Shooting from the hip – or, in business, the all too common shooting from the lip – without collecting and analysing the facts and data isn't the best approach to solving complex business problems. Lean Six Sigma involves understanding what the problem is and then going through several steps to gain a better understanding of it (Define, Measure); working down to the root causes (Analyse); looking at the various solution options and then choosing the most appropriate (Improve); and implementing the solution and holding the gains (Control).

Although this process sounds straightforward and sensible, for many business executives, who believe they know best, it's counter-intuitive.

Consider the vast number of decisions made each year about IT systems, call-centre outsourcing, business re-organisation, new products, and company-wide training programmes. These decisions are often 'solutions' agreed in busy executive meetings – but companies often discover six months or a year later that such solutions do anything but meet requirements or run to budget.

Unfortunately, the very organisations that might benefit most from a Lean Six Sigma-style approach are the least likely to adopt it.

Coming Down with Analysis Paralysis

Getting the balance right is important. During the Analyse phase you may be tempted to get further and further into root cause analysis and lose sight of the primary reason for the improvement project – to make a difference and see positive changes in your business. Your team may get bogged down in the sheer volume of analysis options that can be carried out as they make more discoveries.

Restricting the scope of the project is important to avoid going off on tangents. By all means log those potential opportunities for future projects, but for now stick to the original scope.

Knowing when to end the analysis and start the improve phase can be difficult. Try regarding this decision as a judicial case and weigh up the balance of evidence for and against the 'defendants' – the causes of the problem in your Lean Six Sigma project.

You're probably ready to move to the improve stage if you answer 'yes' to the question:

> 'Are we sure that we understand enough about the process, problem, and causes to develop effective solutions?'

and 'no' to this one:

> 'Is the value of additional data worth the extra cost in time, resources, and momentum?'

The project champion (discussed in Chapter 13) has a key role in ensuring that you keep the business interests at the forefront when answering these questions and in steering the team ahead on the business track.

Achieving Six Sigma – 3.4 defects per 1 million opportunities – may be an aspiration (refer to Chapter 1 for calculating Sigma values), but you won't achieve it in one project. Moving from 2 Sigma to 3 Sigma and then onto 4 Sigma in your Lean Six Sigma projects is entirely normal. Small bite-sized projects move your performance in the right direction, so be prepared to accept just a small increase in the Sigma value of your processes.

Falling into Common Project Traps

Want to know how to ensure project failure? Try a negative brainstorming technique: it's a great icebreaker at the start of a project and is certain to get your team bouncing with ideas. Instead of brainstorming ideas to make the project a success, you brainstorm the opposite: 'How can we ensure project disaster?' You'll be amazed at the number of suggestions the team will have! Then you can turn these negative thoughts into positive ones. You'll end up with a really positive set of suggestions based on practical experience of what can make projects fail.

For starters, here are some common project traps:

Methodology madness

- Not using a structured and planned approach
- Predetermining your solution
- Giving poorly managed handovers
- Allowing the control phase to be weak, so failing to hold the gain

Scope scandals

- Running too many projects at the same time
- Undertaking too large a project
- Having a goal that isn't measurable or is too vague
- Ignoring 'outside-in' customer focus
- Failing to link the goal to a real business need
- Allowing the project scope to keep growing

Team turmoil

- Creating a team with the wrong mix of skills or functional representation (for example, not getting Finance or HR in when needed)
- Offering inadequate training
- Making a poor choice of team leader
- Failing to agree on the time requirements of the team
- Having no shared vision of success

Lack of support

- Using unsupportive key stakeholders
- Having no active project sponsor or champion
- Running competing projects or projects with conflicting objectives
- Allowing poor leadership behaviour
- Failing to allow enough time to run the project systematically

Stifling the Programme Before You've Started

Chances are, some people in your organisation don't share your vision and are all too keen to stamp on your programme before you get it off the ground. Here are a few comments we've heard people say when stifling a Lean Six Sigma programme:

- 'This is just common sense.'
- 'Our place is different.'
- 'It costs too much.'
- 'We're all too busy to do that.'
- 'Let's get back to reality.'
- 'Why change? It's still okay.'
- 'We're not ready for this yet.'
- 'It's a good thought but highly impractical.'
- 'Not that crazy idea again!'

- 'We've always done it this way.'
- 'We're no worse than our competitors!'

Chapter 12 covers dealing with the people aspects of Lean Six Sigma. Win over the doubters in your organisation and you're halfway to making your project succeed.

Ignoring the Soft Stuff

Many traditional Six Sigma training courses cover the 'hard stuff', such as the statistical techniques, the DMAIC methodology, and an extensive array of tools and techniques, but don't deal with the softer tools – the people issues – that you need to gain buy-in and overcome resistance.

Consequently, many novice Six Sigma practitioners try to run projects that focus on statistical tools and blind people with newly learned expressions – and then they're disappointed when their managers or operational workers don't accept the idea.

The *quality* (Q) of the solution that comes from the use of the hard tools and the *acceptance* (A) of the solution that comes from the soft tools are equally important. You need both quality and acceptance to win support and achieve an *effective* (E) outcome.

From our experience in countless projects, the really hard stuff is the soft stuff!

Getting Complacent

Underestimating the amount of energy you need to make your Lean Six Sigma a success is a major pitfall. Complacency sets in surprisingly quickly if you don't drive and lead your programme with urgency. You need an active Lean Six Sigma programme manager, with support from a senior executive, to keep your programme alive, relevant, and on the business agenda.

Changes in organisation are frequent in many businesses. We've seen Lean Six Sigma programmes wither when a programme manager is diverted to internal organisational politics. Ensuring that the senior executive team is actively involved is important. Institutionalising the whole approach is key, so that Lean Six Sigma becomes part of the 'way we do things around here'.

Thinking 'We're Already Doing It'

A quick skim through the Lean Six Sigma literature or rapid overview of your processes may lead you to believe that your organisation's 'already doing it'. Many managers think that they already solve problems using a systematic problem-solving process – but often they don't think about or test solution options properly before putting them into action.

You may think, 'We already use process flowcharts'. Many organisations do use this technique, but often without first understanding the true requirements of the process from the customer's perspective. Unless you adopt Lean Six Sigma in a structured way, you won't be able to fully utilise the power of process mapping techniques to really get under the surface of how your existing processes work.

Genuine senior management buy-in for this kind of peripheral process mapping activity is also unusual. Isolated cases do exist in organisations as part of a cottage industry of enthusiasts who are doing their best but operating outside the scope of a serious senior management-led initiative. A well-designed Lean Six Sigma programme builds on existing knowledge and legitimises improvement work into a framework that involves everyone and introduces a common set of tools across the organisation.

Believing the Myths

A whole series of myths have developed around the use of Lean Six Sigma. For Lean Six Sigma to work in practice, you need to dispel the following ideas:

- **Lean Six Sigma is all we need.** No – Lean Six Sigma can, and should, be integrated with other approaches.
- **Lean Six Sigma is just for manufacturing or production improvement.** No – all processes can be improved. Lean Six Sigma has been used successfully in transaction and service processes.
- **Lean Six Sigma is just about statistical tools and measures.** No – Lean Six Sigma actually involves cultural change.
- **Lean Six Sigma is just about individuals and experts.** No – to work best, Lean Six Sigma involves everyone in a team effort, including senior executives.

Doing the Wrong Things Right

Most people want to do the right things right. Process analysis is a great tool to show you what you're doing in practice and help you answer the questions 'Why?' and 'Are we doing this step correctly?'

According to systems theorist, Russ Ackoff:

> *The righter you do the wrong thing, the wronger you become; if you make a mistake doing the wrong things and correct it, you become wronger; if you make a mistake doing the right thing and correct it, you become righter, therefore it is better to do the right thing wrong than the wrong thing right.*

In fact, you have four options:

1. **Doing the right thing right – most people want to do this.**

   Serving great food and providing top-notch service in a stylish restaurant is an example.

2. **Doing the right thing wrong – apply the tools to fix the problem.**

   Imagine great service in a beautiful restaurant but really bad food.

   Listen to the voice of the customer, recognise that the poor quality of the food is the key driver of customer dissatisfaction, and tackle the root causes of the problem. That is, you analyse the process, discover the critical factors underlying the causes of the problem, and solve those. Often problems can be resolved simply; in this case maybe by using less salt.

3. **Doing the wrong thing right – this is non-value-adding.**

   To continue the example, you concentrate on making the restaurant look even better but still serve awful food. That is, you don't find out what the real customer requirements are, jump to the wrong solution, and spend unnecessary money.

4. **Doing the wrong thing wrong – working to get this right is pointless.**

   For example, spending lots of money restyling the restaurant when customers actually liked the earlier style.

The ultimate danger is kicking off a Lean Six Sigma project to fix the situation in option 4 above – and still end up doing the wrong thing!

Overtraining

Clearly, becoming trained in Lean Six Sigma is important and a well thought out training plan needs to form part of the overall deployment programme. But training works best when it's delivered 'just in time' and at the right level.

In the early days, some organisations undertook company-wide, large scale implementations of Six Sigma and forced hundreds of people onto unnecessary 20-day Black Belt training, which resulted in putting many of them off the whole approach (refer to Chapter 2 for an explanation of the martial arts analogy in Lean Six Sigma).

An organisation just starting to use Lean Six Sigma will be full of opportunities for process improvement that can be tackled using the tools learned on a good foundation Green Belt course. Ideally, this six-day training can be split into three smaller modules of two days each, wrapped around a real project being carried out to ensure the training is delivered at the right level and at the right time to fit into the life of the project.

Avoid the pitfall of believing that Black Belt training must be better than Green Belt training and sending people on a full Black Belt course, complete with advanced statistical training, before starting any projects. Start simply, develop the basic skills, run initial projects quickly to deliver tangible benefits to the organisation, and then select the right candidates to be trained in the Lean Six Sigma advanced tools when needed.

Training people at an advanced level too soon is a waste of money and deters people from using the approach.

Chapter 15

Ten Places to Go for Help

In This Chapter

- Tapping into a wealth of knowledge and experience
- Using the power of the web
- Considering software applications
- Joining a community of interest and developing a network

Sometimes Lean Six Sigma seems a bit daunting. But don't worry, plenty of help and good experience exists if you know where to look and in this chapter we show you where and how to find out what works.

Your Colleagues

A well-managed Lean Six Sigma programme relies on teamwork and support being available for everyone involved across the organisation through an internal network. Support can be offered through a spectrum of different coloured 'belts'; for example, Black Belts supporting Green Belts. Ideally, Black Belts are able to call on support from Master Black Belts who are professional experts in Lean Six Sigma, but in smaller organisations this support maybe outsourced to a specialist.

The 'belt' terminology isn't mandatory. Many organisations just use terms such as 'practitioner' and 'expert' instead of Green Belt and Black Belt. (Chapter 2 explains the martial arts analogy in more detail.)

Being able to access this kind of support network is important. There's a big difference between using a tool in a training environment and operating in the real world, where your first port of call for help is usually your own colleagues.

Your Champion

Every project deserves a good sponsor, or 'champion' (described in detail in Chapter 12). When things get tough, as most projects do from time to time, your project champion is a good source of help. Your champion supports your project team, helps unblock project barriers, and assists you when you need buy-in at a more senior level in your organisation.

Other Organisations

Every year, the number of organisations deploying Lean Six Sigma increases. Over time, the combination of tools and techniques may have changed, but the essentials of using a systematic method, focusing on understanding customer requirements, and improving processes are well tried and tested.

Visiting some other organisations and learning from their experiences is very worthwhile. You may not be able to look deep inside your competitors' businesses, but you can discover lots by visiting similar-sized companies in different sectors. Industry and government special interest groups are a good source of help and often arrange visits for groups to observe companies at work. If you have the chance to visit a Toyota plant, in just a few hours you'll learn a lot about the approach that forms the basis for lean thinking in general.

The Internet

Lots of sites are aimed at Lean Six Sigma devotees. Just for fun, here's a snippet of trivia: if you search the web using the expression 'Six Sigma Pink Floyd', you discover that Roger Waters set up a band in 1964 called Sigma Six before forming Pink Floyd a year later. That was 20 years before Motorola came up with the idea of Six Sigma – progressive rock indeed!

Following are some of our favourite websites devoted to Lean Six Sigma:

- **www.asq.org** The site for The American Society for Quality with very comprehensive online resources and publications.
- **www.catalystconsulting.co.uk** and **www.businessimprovementzone.com** The authors' own websites, regularly updated with new articles and with access to an online learning resource area.

A note about search engines

The Internet's awash with useful information, articles, and guides – if only you could find what you want!

Search engines can speed up your research or investigation – on the web you may even find details of a Lean Six Sigma project similar to one you're working on.

Google is a very useful resource for searching on specifics and is the most popular search engine by far. Others worth trying are Alta Vista, which sometimes goes deeper than Google. Searchy.co.uk is capable of exploring 15 UK based search engines simultaneously. Mamma.com is also a favourite.

- **`www.efqm.org`** Full of useful material and a link into Excellence One, the knowledge library of the European Foundation for Quality Management – essential for anyone serious about learning more about developing quality and excellence across an entire organisation.
- **`www.isssp.com`** Dedicated to Six Sigma, with plenty of articles.
- **`www.isixsigma.com`** The number one (US-focused) Six Sigma website, with bulletin boards, job ads, and links – for addicts only.
- **`www.onesixsigma.com/lean-six-sigma-knowledge-centre/six-sigma-lean-articles-library`** A growing set of new articles on all aspects of Lean and Six Sigma.
- **`www.qualitydigest.com`** A useful online magazine on quality.
- **`www.qfdi.org`** The site for the Quality Function Deployment (QFD) Institute. QFD is an approach to really understand customer requirements and link these to processes, products, and services, and is often used when Lean Six Sigma companies want to design new products and services. QFD is an additional tool used in Design for Six Sigma (DFSS).

Networks and Associations

You can find all sorts of networks and associations relating to Lean Six Sigma. Some networks offer online and offline services to encourage collaboration and knowledge exchange between members, and often hold regular members' meetings.

For example, i&i is a European community of practice for business improvement and innovation. To avoid any 'selling' connotations, this network doesn't permit consultancy organisations to become members.

National and regional quality associations such as the American Society for Quality (ASQ), European Foundation for Quality Management (EFQM), and the British Quality Foundation (BQF) provide opportunities to share good, and not so good, practice through meetings, visits to businesses, conferences, workshops, and online resources, although these aren't dedicated purely to Lean Six Sigma. The EFQM provides members with access to 'Excellence One', an extensive knowledge library offering insights into the approaches used in different organisations.

Conferences

Lean Six Sigma conferences are a regular feature of the conference calendar these days. Conference organisers hold Lean Six Sigma summits every year at different locations around the world. These summits provide a range of mainstream speakers, smaller workshops, networking, and informal discussions regarding every aspect of Lean Six Sigma. Whether you're just starting out or want to keep up with the latest thinking and new developments, these summits are a great source of information.

Use the Internet to stay up-to-date with the conference calendar; `www.onesixsigma.com` has a current diary of conferences and meetings.

Books

You can find a wealth of books on the individual aspects of Lean and Six Sigma, and a few on Lean Six Sigma. Here are a few of our favourites:

- *Implementing Six Sigma* by Forrest Breyfogle III (Wiley-Interscience): Comprehensive reference textbook.
- *Making Six Sigma Last* by George Eckes (Wiley): Cultural aspects of making it happen and succeed.
- *Quantitative Approaches in Business Studies*, 6th edition, by Clare Morris (FT Prentice Hall): An academic textbook, offering a good foundation in statistical methods in business.
- *SPC in the Office* by Mal Owen and John Morgan (Greenfield): Full of useful case studies about using control charts in the office.

- *The Lean Six Sigma Improvement Journey* by John Morgan: Light-hearted coverage of each tool (of which there are many), with the aid of colour-coded illustrations.
- *The Six Sigma Revolution* by George Eckes (Wiley): The principles of Six Sigma.
- *The Six Sigma Way* by Peter Pande, Robert Neuman, and Roland Cavanagh (McGraw-Hill): Good general overview and 'how to'.
- *The Six Sigma Way Team Fieldbook* by Peter Pande, Robert Neuman, and Roland Cavanagh (McGraw-Hill): Practical implementation guide.

Periodicals

Several journals are devoted to Lean and Six Sigma, including:

- *International Journal of Six Sigma and Competitive Advantage* – keeps at the forefront of Six Sigma developments.
- *iSixSigma Magazine* – for Six Sigma professionals, with specialist features on all aspects of the approach; also available online.
- *Quality World* – the magazine of the Chartered Quality Institute in the UK, with regular features on Lean Six Sigma.
- *Six Sigma Forum* – a specialist magazine of the American Society for Quality (ASQ).
- *UK Excellence* – the magazine of the British Quality Foundation, with regular features on Lean Six Sigma.

Software

You can certainly start down the Lean Six Sigma road without having to invest in specialist software, but as your journey proceeds you may want to enhance your toolkit with statistical and other software. In this section, we mention a few of our essentials.

One area of Lean Six Sigma where we recommend *not* using software, especially when starting out, is value stream mapping and process deployment flow-charting. For this, we suggest that you map the process using sticky notes, a pencil, and a large piece of paper pinned to the wall. (Chapter 5 explains this process in more detail.)

That said, if you do decide to use software for process flowcharting, consider Visio, iGraphix, or FlowMap.

Statistical analysis

Most everyday mortals use only a fraction of the full capability of their spreadsheet program, such as Excel. These programs are good at statistical analysis – but because they weren't designed specifically for this purpose, producing even the most basic Pareto chart without help from a kind soul who's produced a template for this purpose is surprisingly challenging.

Fortunately you can find several plug-ins for your spreadsheet program to help you perform Pareto analysis, and slice and dice your data quickly and easily without having to design your own template:

- For more complex statistical analysis, try the Excel plug-in SigmaXL, which lets you produce a variety of displays including SIPOCs, cause and effect diagrams, failure mode and effects analysis, and several types of control chart.
- Most Black Belts and Master Black Belts favour Minitab® Statistical Software. This package has been around for many years and is a favourite of universities and colleges teaching statistics. Minitab is a very comprehensive statistical analysis package designed for serious statistical analysis. Don't try it at home without some serious training as part of an Advanced Green Belt or full Black Belt course.
- JMP Statistical Discovery Software is another package gaining in popularity for use in the world of Lean Six Sigma. It links statistics to a highly visual graphic representation, enabling you to visually explore the relationships between process inputs and outputs, and then to identify the key process variables.
- For more advanced statistical and predictive modelling, take a look at Crystal Ball from Oracle. This popular bit of software is good for forecasting, simulation, and evaluating optimisation options.

Deployment management

For large-scale deployments, consider forming a project library and use tracking software to help you and your colleagues across the organisation manage and report on projects. Software packages such as those from i-Nexus and Instantis are designed specifically for this purpose, and are well worth investigating as your deployment grows across the organisation.

Training and Consultancy Companies

A wide range of specialist training and consulting companies provide services for clients in the Lean Six Sigma arena. In your quest for training, you can find a few global players and lots of smaller specialists.

When you choose a supplier, use the quality × acceptance equation that we describe in Chapter 12. You want your trainer to have excellent technical skills, but also consider how well they could work with your organisation. Will your organisation's culture accept the trainer? Will the trainer instil confidence and provide all the services you require?

In our experience, few organisations bother to check suppliers' references. But unlike choosing a partner or spouse, in business, asking previous clients how well the partnership worked is fine! Working over a long period with a training and consulting company is a bit like a marriage – shared values are a good foundation for belief, integrity, respect, trust, and honesty.

Index

• Numerics •

• A •

• B •

• C •

• E •

• F •

• J •

• K •

• L •

• M •

• N •

• P •

• Q •

• R •

• S •

• V •

• W •

• X •

• Y •

Notes

Notes

Notes

Notes

Notes

Notes

Notes

Notes

Notes